Thought and Object

THOUGHT AND OBJECT

Essays on Intentionality

EDITED BY

ANDREW WOODFIELD

CLARENDON PRESS · OXFORD

Oxford University Press, Walton Street, Oxford OX2 6DP

London Glasgow New York Toronto
Delhi Bombay Calcutta Madras Karachi
Kuala Lumpur Singapore Hong Kong Tokyo
Nairobi Dar es Salaam Cape Town
Melbourne Auckland

and associated companies in
Beirut Berlin Ibadan Mexico City Nicosia

Published in the United States by
Oxford University Press, New York

British Library Cataloguing in Publication Data

Thought and object.
1. Intentionalism
I. Woodfield, Andrew
121   BD143

ISBN 0-19-824606-4
0-19-824677-3 Pbk

First Published 1982
Reprinted 1984

Typeset by CCC, printed and bound in Great Britain by
William Clowes Limited, Beccles and London

Foreword

Questions about the relations between *thoughts* and *things thought about* have long been part of the staple fare of philosophy. In the past decade, however, principally through the influence of Saul Kripke and Hilary Putnam, the notion of *de re* thought has become the focus of new attention. The definition of this notion is itself a matter of controversy, but on one view, a view inspired by Kripke's and Putnam's theories of reference and essence, a *de re* thought has the following two features: it is about an actually existing object, and it is tied to that object *constitutively*, so that the thought could not exist without that very object's existing. The kind of impossibility alluded to is logical or metaphysical rather than causal. The thought could not exist without the object because it is *individuated* in a way that makes its relatedness to that object essential to its nature.

The claim that some thoughts (and beliefs) are *de re* in this sense is typically supported by arguments that appeal to modal intuitions. One form of argument developed by Putnam invites us to imagine an environment which differs from our own in some specified way. We are then asked to judge, concerning a specified psychological predicate, whether it could truly apply to a subject inhabiting that environment. The imaginary environment is thus a heuristic device for eliciting our views about the conditions of application of the predicate. Such thought-experiments need to be set up very carefully, but as long as people understand exactly what is being asked, their judgements tend to tally. We mostly agree that certain predicates would *not* apply, and the reason appears to be, in Putnam's versions of the experiment, that the predicates in question presuppose the existence of things that are absent from the subject's environment. This fact about the predicates is then taken by the *de re* theorist to show that the psychological states ascribed by those predicates simply cannot *obtain* unless the relevant objects exist.

Various questions may be raised about *de re* thought, so conceived. One query concerns the cogency of this form of argument. Does it really show that some mental states are intrinsically *de re*, or does it indicate rather that some of our *ascriptions* are *de re* (i.e. existentially committed)? If the latter only, the argument would not be very exciting, for it is well-known that names and descriptions can occur referentially in 'that'-clauses complementing psychological verbs. An ascriber might say, for example, 'Tom believes that Vienna is impressive' with the intention of referring to the capital of Austria. The speaker presupposes the existence of Vienna, but he does not necessarily imply that the existence of Vienna is required in order for Tom's belief to have the psychological character it has. If it is true that Tom's belief could be the same even if the real Vienna did not exist, then Tom's belief is not *intrinsically de re*. (I should point out that not all writers use the term '*de re*' in this way. Unfortunately, the term is a major source of confusion.)

However, Putnam's argument *is* exciting, because it is designed to apply to ascriptions in which proper names and kind-names occur *opaquely*, as contributing to the characterization of the contents of the mental states in question. Suppose we interpret the same sentence in such a way as to imply that Tom conceives of the city *as* Vienna—that is, he possesses an individual Vienna-concept, a mental equivalent of the name 'Vienna'. Suppose also that the ascription is true on that interpretation. One might say that for Tom's concept to be a genuine Vienna-concept, it must be the product of a causal process traceable back to events in which Vienna itself played a part. It is therefore impossible for Tom to have an individual concept of Vienna unless Vienna really exists or has existed. Any belief in which such a concept figures should be viewed as a historical product whose content is partially constituted by its link with that actual city. The argument passes from certain assumptions of a linguistic kind (e.g. about the semantics of names, and about what it is for a term to occur opaquely in oratio obliqua), to a substantive conclusion about the nature of certain psychological states. But these assumptions are controversial and theory-laden; the argument certainly merits much more careful, detailed study.

Another way of approaching the matter is to grant, for the time being, that we do individuate some thoughts in a *de re* way, and then to raise the question: do we need to do this, or is there always another mode of individuation available whereby the connection between the thought and the object remains inessential? Descartes, Locke, Hume, and indeed the majority of great philosophers, appear to have held that the specific nature of a thought or belief is fixed by its subjective content, so that it is always a separate matter whether a thought with given content corresponds to a particular thing in external reality. The possibility that there might be exceptions to this is hardly ever considered. According to the conception of mind which prevails in our culture (a conception often called 'Cartesian' although it is actually far older than Descartes), thoughts are by nature private to the person experiencing them; so there must be a way of individuating them which respects that privacy.

Nevertheless, the fact that most philosophers have adopted this view does not necessarily mean it is the best view. The *de re* theorist can insinuate doubt in various ways. For example, he may agree that the content of a thought is constitutive, but then argue that sometimes the object is constitutive too. Alternatively he might argue that, where genuinely *de re* thoughts are concerned, the object is *part* of the content. To resolve the issue one must examine the notion of *content* more closely. It is certainly very unclear what philosophers mean when they say that thoughts, beliefs and other mental states have 'intentional content(s)', and it is quite on the cards that they would be unable to give any satisfactory account of it. Perhaps the greatest efforts in this direction were made by Brentano, Husserl, Meinong and Frege. Unfortunately, however, no agreed conclusions were reached.

The problems of intentionality are notoriously puzzling and complicated. No doubt this fact alone accounts for part of their appeal to contemporary philosophers, who relish the challenge of sorting everything out. But there is also a very practical reason why this area is exciting, and that is that it appears to have direct and dramatic implications for *psychology*. If some mental states are indeed irreducibly *de re*, this fact forces us to rethink our conception of the mind and the relation between mind and body.

One doctrine that will need to be dropped is the Cartesian

view that the subject is the best authority on what he or she is thinking. The doctrine has been attacked before, by behaviourists for example, but the present line of attack is more subtle. According to a *de re* theorist, the subject can have full conscious access to the internal subjective aspects of a thought while remaining ignorant about which thought it is. This is because a *de re* thought also has an *external* aspect which consists in its being related to a specific object. Because the external relation is not determined subjectively, the subject is not authoritative about that. A third person might well be in a better position than the subject to know which object the subject is thinking about, hence be better placed in that respect to know which thought it was.

Another interesting consequence is that all versions of the mind-brain identity theory are false. No *de re* mental state about an object that is external to the person's brain can possibly be identical with a state of that brain, since no brain state presupposes the existence of an external object. Any state which did incorporate an environmental object would not be a state of the brain, but would be rather a state of the brain–environment complex.

Not only is there no identifying mind with brain, there is no possibility of empirically *reducing de re* psychological states to physiological states, nor even of effecting a one–one mapping. The physical 'hardware' underlying a *de re* state has two components: the physical state of the organism, and the relevant parts of the world outside. If the latter component were to alter in crucial ways (e.g. if the object of thought were replaced by another object), then the psychological state would be altered too, even though the intra-organismic component remained the same. On this conception the mind is 'wider' than the body; it spreads out on to the environment.

A further consequence is that if a lot of our beliefs, desires, perceptions and attitudes are irreducibly *de re*, then psychologists might as well give up hope of uncovering a system of *laws* linking these types of states to types of behaviour. There are no such laws to be found. We can be sure, in the light of what little we know already about how behaviour is produced, that *de re* states will not figure in the explanatory generalizations of a finished science, since the *de re* psychological taxonomy cuts across functionally relevant boundary lines. Trying to build a science out of states carved up in that way would be like trying to discover how a computer works by looking at the relationship between the computer and the room in which it is kept.

In this short preamble I have not mentioned any of the problems that beset the *de re* theorist when he tries to explicate his conception in more detail. That there are indeed problems will become evident from the essays. My present aim is simply to convey the flavour of what is at stake here. We are confronted with a web of interconnected issues about the individuation of state-types and state-tokens, about intentionality and mental representation, about the semantics of psychological discourse, and also about philosophical methods and techniques. The first three pieces in the collection relate more directly to the '*de re*' question than the second three. But there is no hard division; all six are concerned in one way or another with the prospects for 'solipsistic' psychology—that is, the traditional kind which holds that psychological states do not require the real existence of anything other than the subject.

The philosophy of mind represented here overlaps significantly with the philosophy of language. There are two main reasons why such a rapprochement is called for and why it will inevitably continue to grow. First, there is the recognition that mental content is like semantic content, that mental 'aboutness' is like reference, and that, in general, many of the problems in the two areas have similar contours. In his book *The Language of Thought* (chapter 3), Jerry Fodor claims that 'facts about language will constrain our theories of communication, and theories of communication will in turn constrain our theories about internal representations'. This is surely true, even though the exact nature and extent of the constraints are not yet fully understood.

The second reason is that many of the most fascinating puzzles in the philosophy of mind have to do directly with everyday psychological *idioms* and the curious ways in which they work. Indeed, the whole subject is built upon a realization that philosophers can contribute more by investigating *discourse* about mental states than by investigating the mental states themselves.

This orientation implies that introspection and phenomenological analysis are by and large eschewed; attention is focussed
upon plausible possible speech situations in which one person
talks to another about the mind of a third. The philosopher does
not adopt the role of a participant in the conversations, but is a
theorizer and generalizer about such conversations.

The essays presume some familiarity with existing work in the
field. Two recent contributions stand out: Hilary Putnam's
seminal paper 'The Meaning of "Meaning"', and the book by
Fodor, cited above. Both were published in 1975. Their
importance is evident from the numerous references to them
scattered throughout the collection. Daniel Dennett's paper
'Current Issues in the Philosophy of Mind' (1978) is also worthy
of special mention, being the best general guide to the
background. Perhaps one ought to stress, finally, that the
material in this collection does not purport to be the last word on
the subject. There is a growing body of opinion to the effect that
mentalistic notions like *thinking, believing* and *desiring* belong to a
primitive folk-theory which is either incoherent or false, or in
some other way *flawed*. Further investigation must be carried out
on the intentional idioms before any final conclusions are drawn.

The idea of compiling this volume was stimulated by
conversations I had in 1978 with Dan Dennett and Steve Stich.
During the autumn term of that year I ran a Philosophy of Mind
Workshop, a series of formal and informal discussions conducted
by a dozen philosophers from Bristol University and from the
United States. Much of the workshop was spent in the tentative
exploration of the problems treated here. An early version of
Stich's paper was presented and discussed during the workshop;
Dennett's paper and my own were written shortly after that
fertile period and are, in part at least, reactions to stimuli
encountered then. The papers by Bach, Burge and McGinn have
no connection with the workshop other than the fact that their
topics overlap; they were commissioned and written specifically
for this volume. All six pieces are hitherto unpublished.

I should like to take this opportunity of thanking the United
States–United Kingdom Educational Commission and the
Fulbright–Hays Award Committee, the British Academy, and

the University of Bristol for their generous support of the workshop project. Without their support the workshop could not have taken place; without the stimulation which the workshop provided, this collection would not have come into existence. I should also like to thank the British Academy for funding the preparation of the material for publication.

Bristol   Andrew Woodfield
October 1980

Note

The notes to the text will be found at the end of the chapter to which they refer.

Contents

Contributors

KENT BACH. Professor of Philosophy, San Francisco State University. Co-author (with Robert Harnish) of *Linguistic Communication and Speech Acts* (1979).

TYLER BURGE. Professor of Philosophy, University of California at Los Angeles. Author of numerous journal articles on the philosophy of language and philosophy of mind.

DANIEL C. DENNETT. Professor of Philosophy, Tufts University, Massachusetts. Author of *Content and Consciousness* (1969) and *Brainstorms* (1978).

COLIN McGINN. Lecturer in Philosophy, University College London. Author of numerous journal articles on the philosophy of language and philosophy of mind.

STEPHEN P. STICH. Associate Professor of Philosophy, University of Maryland. Editor of *Innate Ideas* (1975), co-author (with D. A. Jackson) of *The Recombinant DNA Debate* (1978).

ANDREW WOODFIELD. Lecturer in Philosophy, University of Bristol. Author of *Teleology* (1976).

Abstract of
Beyond Belief

Daniel C. Dennett

This paper is a five-part examination of problems with current assumptions about belief.   I. *Introduction.* The doctrine that beliefs are propositional attitudes has wide acceptance, but no stable and received interpretation.   II. *Propositional Attitudes.* Several incompatible doctrines about propositions, and hence about propositional attitudes, have been developed as responses to Frege's demands on propositions or 'Thoughts'. Recently, Putnam and others have presented attacks, with a common theme, on all versions of the standard doctrine.   III. *Sentential Attitudes.* The retreat from these attacks leads (as do other theoretical considerations) to postulating a 'language of thought', but there are serious unsolved problems with this position.   IV. *Notional Attitudes.* An alternative is sketched, midway, in effect, between propositional attitudes and sentential attitudes—neither purely syntactic nor fully semantic. It involves positing a theorist's fiction: the subject's *notional world.*   V. De Re *and* De Dicto *Dismantled.* The foregoing reflections yield alternative diagnoses of the panoply of intuitions conjured up by the *de re/de dicto* literature. These observations grant us the prospect of getting along just fine without anything that could properly be called the distinction between *de re* and *de dicto* beliefs.

Beyond Belief

Daniel C. Dennett

I. Introduction

Suppose we want to talk about beliefs. Why might we want to talk about beliefs? Not just 'because they are there', for it is far from obvious that they *are* there. Beliefs have a less secure position in a critical scientific ontology than, say, electrons or genes, and a less robust presence in the everyday world than, say, toothaches or haircuts. Giving grounds for believing in beliefs is not a gratuitous exercise, but not hopeless either. A plausible and familiar reason for wanting to talk about beliefs would be: because we want to explain and predict human (and animal) behaviour. That is as good a reason as any for wanting to talk about beliefs, but it may not be good enough. It may not be good enough because when one talks about beliefs one implicates oneself in a tangle of philosophical problems from which there may be no escape—save giving up talking about beliefs. In this essay I shall try to untangle, or at least expose, some of these problems, and suggest ways in which we might salvage some theoretically interesting and useful versions of, or substitutes for, the concept of belief. I shall be going over very familiar territory, and virtually everything I shall say has been said before, often by many people, but I think that my particular collection of familiar points, and perhaps the order and importance I give them, will shed some new light on the remarkably resilient puzzles that have grown up in the philosophical literature on belief: puzzles about the content of belief, the nature of belief-states, reference or *aboutness* in belief, and the presumed distinction between *de re* and *de dicto* (or *relational* and *notional*) beliefs. No 'theory' of belief will be explicitly propounded and defended. I have not yet seen what such a philosophical theory would be *for*, which is just as well, since I would be utterly incapable of producing one in any case. Rather, this essay is exploratory and diagnostic, a prelude, with luck, to empirical theories of the phenomena that we now commonly discuss in terms of belief.

If we understand the project to be *getting clear about the concept of belief*, there are still several different ways of conceiving of the project. One way is to consider it a small but important part of the semantics of natural language. Sentences with '. . . believes that . . .' and similar formulae in them are frequently-occurring items of the natural language English, and so one might wish to regiment the presuppositions and implications of their use, much as one would for other natural-language expressions, such as 'yesterday' or 'very' or 'some'. Many of the philosophers contributing to the literature on belief contexts take themselves to be doing just that—just that and nothing more—but typically in the course of explicating their proposed solutions to the familiar puzzles of the semantic theory of 'believes' in English, they advert to doctrines that put them willy-nilly in a different project: defending (that is, defending *as true*) a *psychological* theory (or at least a theory-sketch) of beliefs considered as psychological states. This drift from natural-language semantics (or for that matter, conceptual analysis or ordinary-language philosophy) to meta-theory for psychology is natural if not quite inevitable, traditional if not quite ubiquitous, and even defensible—so long as one recognizes the drift and takes on the additional burdens of argument as one tackles the meta-theoretical problems of psychology. Thus Quine, Putnam, Sellars, Dummett, Fodor and many others have observations to make about how a psychological theory of belief *must go*. There is nothing wrong with drawing such conclusions from one's analysis of the concept of belief, just so long as one remembers that if one's conclusions are sound, and no psychological theory *will* go as one concludes it *must*, the correct further conclusion to draw is: so much the worse for the concept of belief.

If we still want to talk about beliefs (pending such a discovery), we must have some way of picking them out or referring to them or distinguishing them from each other. If beliefs are real—that is, real psychological states of people—there must be indefinitely many ways of referring to them, but for some purposes some ways will be more useful than others. Suppose for instance it was books rather than beliefs that we wished to talk about. One can pick out a book by its title or its text, or by its author, or by its topic, or by the physical location of one of its copies ('the red book on the desk'). What counts as *the same book* depends somewhat on our concerns of the moment: sometimes we mean *the same edition* ('I refer, of course, to the First Folio *Hamlet*'); sometimes we mean merely *the same text* (modulo a few errors or corrections); sometimes merely the same text or a good translation of it (otherwise how many of us could claim to have read any books by Tolstoy?). This last conception of a book is in some regards privileged: it is *what we usually mean* when we talk—without special provisos or contextual cues—about the books we have read or written or want to buy. We use 'a copy of . . .' to prefix the notion of a book in more concrete transactions. You might refute your opponent with *Word and Object* or, failing that, hit him over the head with *a copy of . . . Being and Time.*

There is similar room for variation in talking about beliefs, but the privileged way of referring to beliefs, what we usually mean and are taken to mean in the absence of special provisos or contextual cues, is *the proposition believed*: e.g., the belief that snow is white, which is *the same belief* when believed by Tom, Dick, and Harry, and also when believed by monolingual Frenchmen—though the particular belief-*tokens* in Tom, Dick, Harry, Alphonse and the rest, like the individual, dog-eared and ink-stained personal copies of a book, might differ in all sorts of ways that were not of interest to us given our normal purposes in talking about beliefs.[1]

Beliefs, standardly, are viewed as *propositional attitudes*. The term is Russell's (1940). There are three degrees of freedom in the formulae of propositional attitude: person, attitude type, proposition; x believes that p, or y believes that p; x *believes* that p or *fears* that p or *hopes* that p; x believes that p or that q, and so forth. So we can speak of a person believing many different propositions, or of a proposition being believed by many different people, or even of a proposition being variously 'taken' by different people, or by the same person at different times—I used to doubt that *p*, but now I am certain that *p*. There are other ways of referring to beliefs, such as 'the belief that made Mary blush', or 'McCarthy's most controversial belief', but these are parasitic; one can go on to ask after such a reference: 'and which belief is that?' in hopes of getting an *identifying* reference—e.g., 'Mary's belief that Tom knew her secret'. Some day (some people think) we will be able to identify beliefs neurophysiologically ('the belief in Tom's cortex with physical feature *F*'), but for the present, at least, we have no way of singling out a belief the way we can single out a book by a physical description of one of its copies or tokens.

The orthodoxy of the view that beliefs are propositional attitudes persists in spite of a host of problems. There have always been relatively 'pure' *philosophers'* problems, about the metaphysical status and identity conditions for propositions, for instance, but with the newfound interest in *cognitive science*, there must now be added the *psychologists'* problems, about the conditions for individual instantiation of belief-states, for instance. The attempt to harness propositional attitude talk as a descriptive medium for empirical theory in psychology puts a salutary strain on the orthodox assumptions, and is prompting a rethinking by philosophers. In the nick of time, one might add, since it is distinctly unsettling to observe the enthusiasm with which non-philosophers in cognitive science are now adopting propositional attitude formulations for their own purposes, in the innocent belief that any concept so popular among philosophers must be sound, agreed upon, and well tested. If only it were so.

II. Propositional Attitudes
If we think that a good way of characterizing a person's psychological state is characterizing that person's propositional attitudes, then we must suppose that a critical requirement for getting the *right* psychological description of a person will be specifying the *right* propositions for those attitudes. This in turn requires us to make up our minds about what a proposition is, and more important, to have some stable view about what counts as two different propositions, and what counts as one. But there is no consensus on these utterly fundamental matters. In fact,

there are three quite different general characterizations of propositions in the literature.

(1) Propositions are *sentence-like entities*, constructed from parts according to a syntax. Like sentences, propositions admit of a type–token distinction; proposition tokens of the same proposition type are to be found in the minds (or brains) or believers of the same belief. It must be this view of propositions that is being appealed to in the debate among 'cognitive scientists' about *forms of mental representation*: is all mental representation propositional, or is some imagistic or analogue? Among philosophers, Harman (1973, 1977) is most explicit in the expression of this view of propositions.

(2) Propositions are *sets of possible worlds*. Two sentences express the same proposition just in case they are true in exactly the same set of possible worlds. On this view propositions themselves have no syntactic properties, and one cannot speak of their having instances or tokens in a brain or mind, or on a page. Stalnaker (1976) defends this view of propositions. See also Field (1978) and Lewis (1979) for other good discussions of this oft-discussed view.

(3) Propositions are something like *collections or arrangements of objects and properties in the world*. The proposition that Tom is tall consists of Tom (himself) and the attribute or property tallness. Russell held such a view, and Donnellan (1974), Kaplan (1973, 1978, forthcoming), and Perry (1977, 1979) have recently defended special versions of it. Echoes of the theme can be heard in many different places—for instance in correspondence theories of truth that claim that what makes a sentence true is correspondence with a fact 'in the world'—where a fact turns out to be a true proposition.

What seems to me to unite this disparate group of views about what propositions *are* is a set of three classical demands about what propositions must *do* in a theory. The three demands are due to Frege, whose notion of a *Thought* is the backbone of the currently orthodox understanding of propositions (see Perry 1977). And the diversity of doctrine about propositions is due to the fact that these three demands cannot be simultaneously satisfied, as we shall see. According to the Fregean view, a proposition (a Fregean *Thought*) must have three defining characteristics:

> (a) It is a (final, constant, underived) *truth-value-bearer*. If p is true and q is false, p and q are not the same proposition. (Cf. Stich 1978b: 'If a pair of states can be type identical . . . while differing in truth value, than the states are not beliefs as we ordinarily conceive of them.') (Cf. also Fodor 1980.)

This condition is required by the common view of propositions as the ultimate medium of information transfer. If I know *something* and communicate *it* to you (in English, or French, or by a gesture or by drawing a picture) what you acquire is *the proposition* I know. One can befriend condition (a) without subscribing to this view of communication or information transfer, however. Evans (1980) is an example.

> (b) It is composed of *intensions*, where intensions are understood *à la* Carnap as extension-determiners. Different intensions can determine the same extension: the intension of 'three squared' is not the intension of 'the number of planets' but both determine the same extension. Different extensions cannot be determined by one intension, however.

To say that intensions determine extensions is not to say that intensions are *means* or *methods* of *figuring out* extensions. Evans, in lectures in Oxford in 1979, has drawn attention to the tendency to slide into this understanding of intensions, and has suggested that it contributes plausibility to views such as Dummett's (1973, 1975) to the effect that what one knows when one knows meanings (or intensions) is something like a *route* or *method of verification* or *procedure*. Whether this plausibility is entirely spurious is an open, and important, question.

Condition (b) cuts two ways. First, since the intensions of which a proposition is composed fix their extensions in the world, *what a proposition is about* is one of its defining characteristics. If p is about a and q is not about a, p and q are not the same proposition. Second, since extension does not determine intension, the fact that p and q are both about a, and both attribute F to a, does not suffice to show that p and q are the same proposition; p and q may be about a 'in different ways'—they may refer to a via different intensions.

(For Frege, conditions (a) and (b) were unified by his doctrine that a declarative sentence in its entirety had an extension: either the True or the False. In other words, the intensional whole, the Thought, determines an extension just as its parts do, but whereas its parts determine objects or sets as extensions, the Thought itself has as its extension either the True or the False.)

(c) It is 'graspable' by the mind.

Frege does not tell us anything about what grasping a Thought consists in, and has often been criticized for this. What mysterious sort of transaction between the mind (or brain) and an abstract, Platonic object—the Thought—is this supposed to be? (See, e.g., Fodor 1975, 1980; Field 1978; Harman 1977.) This question invites an excursion into heavy-duty metaphysics and speculative psychology, but that excursion can be postponed by noting, as Churchland (1979) urges, that the catalogue of propositional attitude predicates bears a tempting analogy with the catalogue of physical measure predicates. A small sample:

. . . believes that p	. . . has a length of metres of n
. . . desires that p	. . . has a volume in metres^3 of n
. . . suspects that p	. . . has a velocity in m/sec of n
. . . is thinking that p	. . . has a temperature in degrees K of n

Churchland's suggestion is that the metaphysical implications, if any, of the propositional attitude predicates are the same as those of the physical measure predicates:

The idea that believing that p is a matter of standing in some appropriate relation to an abstract entity (the proposition that p) seem to me to have nothing more to recommend it than would the parallel suggestion that weighing 5 kg is at bottom a matter of standing in some suitable relation to an abstract entity (the number 5). For contexts of this latter kind, at least, the relational construal is highly procrustean. Contexts like

> *x* weighs 5 kg
> *x* moves at 5 m/sec
> *x* radiates at 5 joules/sec

are more plausibly catalogued with contexts like

> *x* weighs very little
> *x* moves quickly
> *x* radiates copiously.

> In the latter three cases, what follows the main verb has a transparently *adverbial* function. The same adverbial function, I suggest, is being performed in the former cases as well. The only difference is that using singular terms for number in adverbial position provides a more precise, systematic, and useful way of modifying the main verb, especially when said position is open to quantification. [1979, p. 105.]

This construal of propositional attitudes does not by itself dissolve the metaphysical problems about propositions, as we shall see, but by binding their fate to the fate of numbers in physics, it disarms the suspicion that there is a *special* problem about abstract objects in psychology. Moreover, it permits us to distinguish two views that are often run together. It is often thought that *taking propositions seriously* in psychology must involve taking propositions to *play a causal role* of some sort in psychological events. Thus one is led to ask, as Harman (1977) does, about the *function* of propositions *in thought.* For propositions to have such a function, they must be concrete—or have concrete tokens—and this leads one inevitably to a version of view (1): propositions are sentence-like entities. One could not suppose that sets of possible worlds or arrangements of things and properties were themselves 'in the head', and only something in the head could play a causal role in psychology. One might take propositions seriously, however, without committing oneself to this line; one might take them just as seriously as physicists take numbers. On Churchland's construal, the *function* of a proposition is just to be the denotation of a singular term completing the 'adverbial' modifier in a predicate of propositional attitude, a predicate we want to use to characterize someone's thought or belief or other pyschological state. This is *not* the view that takes propositional attitude predicates to have no logical structure; indeed it holds out the promise that the formal relations among propositions, like the formal relations among numbers, can be usefully exploited in forming the predicates of a science. It is the view that propositions are abstract objects useful in 'measuring' the psychological states of creatures. This leaves it open for us to prove or discover later that whenever a creature has a particular propositional attitude, something in the creature mirrors the 'form' of the proposition— e.g., is somehow isomorphic or homomorphic with the (canonically expressed) clause that *expresses* the proposition in the sentence of propositional attitude. One need not, and should not, *presuppose* any version of this strong claim, however, as part of one's initial understanding of the meaning of propositional attitude predicates.[2]

Failure to make this distinction, and stick to it, has created a frustrating communication problem in the literature. In a typical instance, a debate arises about the *form* of the propositions in some special context of propositional attitude: e.g., are the propositions in question universally quantified conditionals, or indefinitely long disjunctions, or is there self-reference within the propositions? What is not made clear is whether the debate is taken to be about the actual form of internal cerebral structures (in which case, for instance, the unwieldiness of indefinitely long disjunctions poses a real problem), or is rather a debate just about the correct logical form of the abstract objects, the propositions that are required to complete the predicates under discussion (in which case infinite disjunctions need pose no more problems than π in physics). Perhaps some debaters are beguiled by a tacit assumption that the issue cannot be real or substantial unless it is an issue *directly* about the physical form of structures ('syntactic' structures) in the brain, but other debaters understand that this is not so, and hence persist on their side of the debate without acknowledging the live possibility that the two sides are talking past each other. To avoid this familiar problem I shall cleave explicitly to the minimal, Churchland interpretation, as a neutral base of operations from which to explore the prospects of the stronger interpretations.

With this metaphysically restrained conception of propositional attitudes in mind, we can then define Frege's elusive notion

of graspability quite straightforwardly: propositions are graspable if and only if predicates of propositional attitude are
projectible, predictive, well-behaved predicates of psychological
theory. (One might in the same spirit say that the success of
physics, with its reliance on numbers as predicate-forming
functors, shows that numbers are graspable by physical objects
and processes!) The rationale for this version of graspability is
that Frege's demand that propositions be something the mind
can grasp is tantamount to the demand that propositions *make a
difference* to a mind; that is to say, to a creature's psychological
state. What a person does is supposed to be a function of his
psychological state; variations in psychological state should
predict variations in behaviour. (That's what a psychological
state is supposed to be: a state, variation in which is critical to
behaviour.)[3] Now if people's psychological states vary directly
with their propositional attitude characterizations so that, for
instance, changing one's propositional attitudes is changing one's
psychological state, and sharing a propositional attitude with
another is being psychologically similar in some way to the
other, then propositions figure systematically in a perspicuous
interpretation of the psychology of people—and a vivid way of
putting this would be to say that people (or dogs or cats, if that
is how the facts turn out) grasp the propositions that figure in the
psychological predicates that apply to them. No more marvellous
sort of 'entertaining' of either abstract objects or their concrete
stand-ins is—*as of yet*—implied by asserting condition (c). Frege
no doubt had something more ambitious in mind, but this weaker
version of graspability is sufficiently demanding to create the
conflict between condition (c) and conditions (a) and (b).

A number of writers have recently offered arguments to show
that conditions (a–c) are not jointly satisfiable; what is graspable
cannot at the same time be an extension-determiner or ultimate
truth-value-bearer: Putnam (1975a); Fodor (1980); Perry
(1977, 1979); Kaplan (forthcoming); Stich (1978b). (Among
the many related discussions, see especially McDowell 1977 and
Burge 1979a.)

First there is Putnam's notorious thought-experiment about
Twin-Earth. Briefly (since we will not pause now to explore the

myriad objections that have been raised), the imagined case is this: there is a planet, Twin-Earth, that is a near duplicate of Earth, right down to containing replicas or *Doppelgängers* of all the people, places, things, events, on Earth. There is one difference: lakes, rivers, clouds, waterpipes, bathtubs, living tissues . . . contain not H_2O but 'XYZ'—something chemically different, but indistinguishable in its normally observable macro-properties, from water, that is to say, H_2O. Twin-Earthians *call* this liquid 'water', of course, being atom-for-atom replicas of us (ignore the high proportion of water molecules in us, for the sake of the argument!). Now since my *Doppelgänger* and I are *physical* replicas (please, for the sake of the argument), we are surely *psychological* replicas as well: we instantiate all the same theories above the level at which H_2O and XYZ are distinguishable. We have all the same psychological states then. But where my beliefs are about water, my *Doppelgänger's* beliefs (though of exactly the same 'shape') are not about water, but about XYZ. We believe different propositions. For instance, the belief I would express with the words 'water is H_2O' is *about water*, and *true*; its counterpart in my *Doppelgänger*, which he would express with just the same sounds, of course, is *not* about water, but about what he calls 'water', namely XYZ, and it is *false*. We are psychological twins, but not propositional attitude twins. Propositional attitudes can vary independently of psychological state, so propositions (understood 'classically' as meeting conditions (a) and (b)) are not graspable. As Putnam puts it, something must give: either meaning 'ain't in the head', or meaning doesn't determine extension.

Stich (1978b) points out that it is instructive to compare this result with a similar but less drastic point often made about the state of *knowing*. It is often remarked that whereas 'believes' is a psychological verb, 'knows' is not—or not purely—a psycholog-ical verb, since *x knows that p* entails the truth of *p*, something that must in general be external to the psychology of *x*. Hence, it is said, while *believing that p* may be considered a pure psychological (or mental) state, *knowing that p* is a 'mongrel' or 'hybrid' state, partly psychological, partly something else—epistemic. In this case it is the verb component of the predicate of propositional attitude that renders the whole predicate psychologically impure and non-projectible. (Note that it *is* non-projectible: simple experiments involving deception or illusion would immediately show that '*x* will press the button when *x* knows that *p*' is a less reliable predictor than, say, '*x* will press the button when *x* is sure that *p*'.) What Putnam's thought-experiment claims to show, however, is that even when the verb is an apparently pure psychological verb, the mere fact that the propositional component (*any* propositional component) must meet conditions (a) and (b) renders the entire predicate psychologically impure.

Putnam's thought-experiment is hardly uncontroversial. As it stands it rests on dubious doctrines of natural kinds and rigid designation, but easy variations on his basic theme can avoid at least some of the most common objections. For instance, suppose Twin-Earth is just like Earth except that my wallet is in my coat pocket and my *Doppelgänger*'s wallet isn't in his coat pocket. I believe (truly) that my wallet is in my coat pocket. My *Doppelgänger* has the counterpart belief. His is false, mine is true; his is not about what mine is about—viz., *my* wallet. Different propositions, different propositional attitudes, same psychology.

In any event, Kaplan (forthcoming) has produced an argument about a similar case, with a similar conclusion, which is perhaps more compelling since it does not rely on going along with outlandish thought-experiments about near-duplicate universes, or on the intuitions about natural kinds Putnam must invoke to support the claim that XYZ is not just 'another kind of water'.

Kaplan quotes Frege:

> If someone wants to say the same today as he expressed yesterday using the word 'today', he must replace this word with 'yesterday'. Although the thought is the same its verbal expression must be different so that the sense, which would otherwise be affected by the differing times of utterance, is readjusted. [Frege 1956.]

But he goes on to note what Frege missed:

> If one says 'Today is beautiful' on Tuesday and 'Yesterday was beautiful' on Wednesday, one expresses the same thought according to the passage quoted. Yet one can clearly lose track of the days and not realize one is expressing the same thought [Fregean Thought,

our proposition]. It seems then that thoughts are not appropriate bearers of cognitive significance.

Perry offers yet another argument, which will be discussed later, and there are still further arguments and persuasions to be found in the literature cited earlier.[4]

I do not want to endorse any of these arguments here and now, but I also want to resist the urge—which few can resist, apparently—to dig the trenches here and now and fight to the death on the terrain provided by reflections on Twin-Earth, Natural Kinds, and What Frege Really Meant. I propose to yield a little ground, and see where we are led.

Suppose these arguments are sound. What is their conclusion? One claim to be extracted from Kaplan (to which Putnam and Perry would assent, I gather) is this: If there is some indexical functor in my thought or belief, such as 'now' or 'today', the proposition I am 'related to'—the proposition that fills the slot in the correct propositional attitude predicate applied to me—can depend crucially (but imperceptibly to me) on such events as the moving of a clock hand at the Greenwich Observatory. But it is frankly incredible to suppose that my psychological state (my behaviour-predicting state) might depend not just on my internal constitution at the time but at the least also on such causally remote features as the disposition of the parts of some official time-keeper. That is not to say that my future behaviour and psychology might not be *indirectly* a function of my actual propositional attitudes on occasion—however unknown to me they were. For instance, if I bet on a horse or deny an accusation under oath, the *long range* effects of this action on me may be more accurately predicted from the proposition I *in fact* expressed (and even believed—see Burge 1979a) than from the proposition I as it were took myself to be expressing—or believing![5] This acknowledgement, however, only heightens the contrast between the *unreliable* or *variable* 'accessibility' of propositions, and the built-in or constitutive accessibility of... *what?* If propositions in the Fregean mould are seen by these arguments to be psychologically inert (at least under certain special circumstances), what is the more accessible, graspable, effective object for the propositional role?

III. Sentential Attitudes

With what shall we replace propositions? The most compelling answer (if only to judge by the number of its sympathizers) is: something like *sentences in the head*. (See, e.g., Fodor 1975, 1980; Field 1978; Kaplan forthcoming; Schiffer 1978; Harman 1977—but for some eminent second thoughts, see Quine 1969.) In the end we shall find this answer unsatisfactory, but understanding its appeal is an essential preliminary, I think, to the task of finding a better retreat from propositions. There are many routes to *sentential attitudes*.

Here is the simplest: when one *figuratively* 'grasps' a proposition, which is an abstract object, one must *literally* grasp something concrete but somehow proposition-like. What could this be but a sentence in the mind or brain—a sentence of Mentalese? (For those who already hold the view that propositions are sentence-like things, this is a short retreat indeed; it consists of giving up conditions (a) and (b) for propositions—but then, if propositions are to be something more than mere uninterpreted sentences, what more are they? Something must be put in place of (a) and (b).)

Here is another route to sentential attitudes. What are the actual *constituents* of belief-states about dogs and cats? Not real live dogs and cats, obviously, but . . . symbols for, or representations of, dogs and cats. The belief that the cat is on the mat consists somehow of a structured representation composed of symbols for the cat and the mat and the *on* relation—a sentence of sorts (or maybe a picture of sorts[6])—to which its contemplator says 'Yes!' It is not that belief doesn't eventually have to tie the believer to the world, to real live dogs and cats, but the problem of that tying relation can be isolated and postponed. It is turned into the apparently more tractable and familiar problem of the *reference* of the terms in sentences—in this case inner mental sentences. One imagines that the great horses of logical instruction—Frege, Carnap, Tarski—can be straightforwardly harnessed for this task. (See Field 1978 for the most explicit defence of this route.)

Here is the third route. We need a physical, causal explanation of the phenomenon of opacity; the fact that believing that it would be nice to marry Jocasta is a state with different psychological consequences, different effects in the world, from the state of believing that it would be nice to marry the mother of Oedipus, in spite of the now well known identity. A tempting suggestion is that these two different states, in their physical realizations in a believer, have in effect a syntax, and the syntax of one state resembles and differs from the syntax of the other in just the ways the two sentences of attribution resemble and differ; and that the different effects of the two states can be traced eventually to these differences in physical structure. This can be viewed as an explanation of the opacity of *indirect* discourse by subsuming it under the *super*-opacity of a bit of *direct* discourse— the strict quotation, in effect, of different Mentalese sentences (Fodor 1980).

Here, finally, is the route to sentential attitudes of most immediate relevance to the problems we have discovered with propositional attitudes. Apparently what makes the Frege–Kaplan 'today' and 'yesterday' example work is what might be called the impermeability of propositions to indexicals. This impermeability is made explicit in Quine's surrogate for propositions, *eternal sentences*, which are equipped whenever needed with bound variables of time and place and person to remove the variable or perspectival effect of indexicals (Quine 1960). A psychologically perspicuous alternative to propositions would resist precisely this move, and somehow build indexical elements in where needed. An obvious model is ready to hand: sentences—ordinary, external, concrete, uttered sentences of natural languages, spoken or written. Sentences are to be understood first and foremost as syntactically individuated objects—as strings of symbols of particular 'shapes'—and it is a standard observation about sentences so individuated that tokens of a particular sentence type may 'express' different propositions depending on 'context'. Tokens of the sentence type 'I am tired' express different propositions in different mouths at different times; sometimes 'I am tired' expresses a true proposition about Jones, and sometimes a false proposition about Smith. Perhaps there are, as Quine claims, some sentence types, the eternal sentences, all of whose tokens in effect express the same proposition. (Quine—no friend of propositions—must be more devious in making this claim.) But it is precisely the power of the other, *non*-eternal sentences to be context-*bound* that is needed, intuitively, for the role of individuating psychologically salient states and events. Indexicality of sentences appears to be the linguistic counterpart of that relativity to a subjective point of view that is a hallmark of mental states. (Castañeda 1966, 1967, 1968; Perry 1977, 1979; Kaplan forthcoming; Lewis 1979.)

If *what a sentence means* were taken to be *the proposition it expresses*, then different tokens of an indexical sentence type will mean different things, and yet there does seem both room and need for a sense of 'meaning' according to which we can say that all tokens of a sentence type *mean the same thing*. A sentence *type*, even an indexical type such as 'I am tired', means something—just one 'thing'—and hence in this sense so do all its tokens. That same thing is no proposition, of course. Call it, Kaplan suggests, the sentence's *character*. 'The character of an expression is set by linguistic conventions and, in turn, determines the content of the expression in every context.' Kaplan's suggestion unfolds into a symmetrical two-stage picture of sentence interpretation:

> Just as it was convenient to represent contents by functions from possible circumstances to extensions (Carnap's intensions), so it is convenient to represent characters by functions from possible contexts of utterance to contents.... This gives us the following picture:
>
> Character: Contexts → Content
> Content: Circumstances → Extensions
>
> or in more familiar language,
>
> Meaning + Context → Intension
> Intension + Possible World → Extension

Although Kaplan is talking about public, external sentences— not sentences in the head, in Mentalese or brain writing—the relevance of this sort of linguistic meaning to psychology is immediately apparent. Kaplan comments: 'Because character is what is set by linguistic conventions, it is natural to think of it as *meaning* in the sense of what is known by the competent language user.' No mastery of my native tongue will ensure that I can tell *what proposition* I have expressed when I utter a sentence, but my competence as a native speaker does, apparently, give me access to the *character* of what I have said. Could we not generalize the point to the postulated language of thought, and treat the character of Mentalese sentences as what is directly grasped when one 'entertains' (mentally 'utters') a Mentalese sentence? Couldn't the *objects* of belief be the characters of Mentalese sentences instead of the propositions they express?

Commenting on Kaplan, Perry (1977, 1979) develops this theme. Where Kaplan speaks of *expressions* of the same *character*, Perry moves the issue inside the mind and speaks of people *entertaining* the same *senses*, and where Kaplan speaks of *expressions* having the same *content*, Perry speaks of people *thinking* the same *thought*. (Perry's terms deliberately echo Frege, of course.) He nicely expresses the appeal of this theoretical move in a passage also cited by Kaplan:

> We use senses [Kaplan's *characters*, in essence] to individuate psychological states in explaining and predicting action. It is the sense entertained, and not the thought apprehended, that is tied to human action. When you and I entertain the sense of 'A bear is about to attack me', we behave similarly. We both roll up in a ball and try to be as still as possible. Different thoughts apprehended, same sense entertained, same behavior. When you and I both apprehend the thought that I am about to be attacked by a bear, we behave differently. I roll up in a ball, you run to get help. Same thought apprehended, different sense entertained, different behavior. Again, when you believe that the meeting begins on a given day at noon by entertaining, the day before, the sense of 'the meeting begins tomorrow at noon', you are idle. Apprehending the same thought the next day, by entertaining the sense of 'the meeting begins now', you jump up from your chair and run down the hall. [1977, p. 494.]

The idea, then, is that we postulate a language of thought, possibly entirely distinct from any natural language a believer may know, and adapt Kaplan's two-stage account of meaning (character + content), which was designed initially for the semantic interpretation of natural-language expressions, as a two-stage account of the semantic interpretation of psychological states. The first stage, Perry's *senses* (modelled on *characters*) would give us psychologically pure predicates, with graspable objects; the second stage, Perry's *thoughts* (modelled on Kaplan's *contents*), would be psychologically impure, but would complete the job of semantic interpretation by taking us all the way (via Carnapian intensions, in effect) to extensions—things in the world for beliefs to be about.

Here is the proposal from another perspective. Suppose we had begun with the question: what is it about a creature (about any entity with psychological states) that determines what it believes? That is, what features of the entity, considered all by itself in isolation from its embedding in the world, fix the propositions of its propositional attitudes? To this question the startling answer from Putnam is: *nothing!* Everything that is true about such an entity considered by itself is *insufficient* to determine its beliefs (its propositional attitudes). Facts about the environmental/causal/historical embedding of the entity—the 'context of utterance' in effect—must be added before we have enough to fix propositions.

Why is the answer startling? To anyone with fond memories of Descartes's *Meditations* it ought to be startling, for in the *Meditations* it seemed unshakeably certain that the *only* matter that was fixed or determined solely within the boundaries of Descartes's mind was exactly which thoughts and beliefs he was having (which propositions he was entertaining). Descartes might bewail his inability to tell which of his beliefs or thoughts were true, which of his perceptions veridical, but which thoughts and beliefs they were, the identity of his own personal candidates for truth and falsehood, seemed to be fully determined by the internal nature of his own mind, and moreover clearly and distinctly graspable by him. If Putnam, Kaplan and Perry are right, however, Descartes was worse off than he thought: he couldn't even be certain which propositions he was entertaining.

There are at least four ways of resolving this conflict. One could side with Descartes and search for a convincing dismissal of the Putnam line of thought. One could accept the Putnamian conclusion and dismiss Descartes. One could note that Putnam's case is not directly an attack on Descartes, since it presupposes the physicality of the mind, which of course Descartes would disavow; one could say that Descartes could grant that everything *physical* about me and my *Doppelgänger* underdetermines our propositional attitude, but they are nevertheless 'internally' determined by features of our non-physical minds—which must be just dissimilar enough in their nature to fix our different propositional attitudes. Or one could try for a more irenic compromise, accepting Putnam's case against propositions understood as objects meeting conditions (a–c), and holding that what Descartes was in a privileged position to grasp were rather the *true* psychological objects of his attitudes, not propositions but Perry's senses.

Following the last course, we change our initial question: what, then, is the *organismic contribution* to the fixation of propositional attitudes? How shall we characterize what we get when we subtract facts about context or embedding from the determining whole? This remainder, however we ought to characterize it, is the proper domain of psychology, 'pure' psychology, or in Putnam's phrase, 'psychology in the narrow sense'. Focussing of the organismic contribution in isolation is what Putnam calls *methodological solipsism*. When Fodor adopts the term and recommends methodological solipsism as a research strategy in cognitive psychology (1980) it is precisely this move he is recommending.

But how does one proceed with this strategy? How shall we characterize the organismic contribution? It ought to be analogous to Kaplan's notion of character, so we start, as Perry did, by psychologizing Kaplan's schema. When we attempt this we notice that Kaplan's schema is incomplete, but can be straightforwardly extended in consonance with his supporting comments. Recall that Kaplan claimed that 'linguistic conventions' determine the character of any particular expression type. So Kaplan's two-stage interpretation process is preceded in effect by an earlier stage (0), governed by linguistic conventions:

(0) Syntactic features + linguistic conventions → Character
(1) Character + Context → Content
(2) Content + Circumstances → Extension

When we psychologize the enlarged schema, what will we place in the first gap? What will be our analogue of the syntactic features of utterances which, given linguistic conventions, fix character? This is where our commitment to a *language of thought* comes in. The 'expressions' in the language of thought are needed as the 'raw material' for psychologico-semantic interpretation of psychological states.

This is just what we expected, of course, and at first everything seems to run nicely. Consider Putnam's thought-experiment. When he introduces a *Doppelgänger* or physical replica, he is tacitly relying on our acquiescence in the claim that since two exact replicas—my *Doppelgänger* and I—have exactly the same structure at all levels of analysis from the microscopic on up, *whatever* syntactically defined systems one of us may embody the other embodies as well. If *I* think in brain writing or Mentalese, thoughts with exactly the same 'shapes' occur in my *Doppelgänger* as well, and the further tacit corollary is that my thoughts and my *Doppelgänger*'s will also be *character*-type-identical, *in virtue of their syntactic type identity. My* thoughts, though, are about *me*, while *his* thoughts are about *him*—even though his names for himself are syntactically the same as my names for myself: we both call ourselves 'I' or 'Dennett'. Putnam's case then apparently nicely illustrates Kaplan's schema in action. My *Doppelgänger* and I have thoughts with the same character (sense, for Perry), and all that is needed is context—Earth or Twin-Earth—to explain the difference in content (proposition, Perry's and Frege's *thought*) expressed, and hence the difference in extension, given circumstances.

What assumptions, though, permit the tacit corollary that syntactic type identity is sufficient for character type identity? Why isn't it possible that although a thought of the *shape* 'I am tired' occurs in both me and my *Doppelgänger*, in *him* the thought with that shape means *snow is white*—and hence differs not only in proposition expressed, but in character as well? I grant that on any sane view of Mentalese (if there is any), character type identity *should* follow from physical replicahood, but why? It must be due to a difference between people and, say, books, for an atom-for-atom replica of *The Autobiography of Malcolm X* on another planet (or just anywhere they speak Schmenglish) might not be an autobiography of anyone; it might be a monograph on epistemic logic or a history of warfare.[7] The reason we can ignore this alternative in the case of Mentalese must have to do with a more intimate relationship between form and function in the case of anything that could pass as Mentalese, in contrast with a natural language. So the role played by linguistic conventions in stage (0) will have to be played in the psychological version of the schema by something that is not in any ordinary sense *conventional* at all.

In fact, when we turn to the attempt to fill in the details of stage (0) for the psychological schema, we uncover a host of perplexities. Whence come the syntactic features of Mentalese? Psychology is *not* literary hermeneutics; the 'text' is not *given*. Which shapes of things in the head count? It appears that Kaplan has skipped over yet another prior stage in his schema.

(− 1) physical features + design considerations (in other
        words: *minus* functional irrelevancies)
                                → syntactic features

Differences in the typeface, colour and size of written tokens, and in the volume, pitch and timbre of spoken tokens don't count as syntactic differences, except when it can be shown that they *function* as syntactic differences by marking combinatorial 'valences', possibilities of meaningful variation, etc. A syntactic characterization is a considerable abstraction from the physical features of tokens; Morse code tokens occurring in time can share their syntax with printed English sentence tokens.

By analogy then we can expect brain writing tokens to differ in many physical features and yet count as sharing a syntax. Our model grants us this elbow room for declaring physically quite different 'systems of representation' to be mere 'notational variants'. This is in any case a familiar idea, being only a special case of the freedom of physical realization championed by functionalist theories of mind (e.g., Putnam 1960, 1975b; Fodor 1975). Surely the believer in sentential attitude psychology will be grateful for this elbow room, for the position beckoning at the end of this trail is startlingly strong. Let us call it *sententialism*:

(S) x believes what y believes if and only if $(\exists L)$ $(\exists s)$ (L is a language of thought and s is a sentence of L and there is a token of s in x and a token of s in y)

(We must understand that these tokens have to be in the functionally relevant and similar places, of course. One can't come to believe what Jones believes by writing his beliefs (in L) on slips of paper and swallowing them.)

On this view we must *share* a language of thought to believe *the same thing*—though we no longer mean by 'the same thing' the same proposition. The idea is that from the point of view of psychology a different type-casting is more appropriate, according to which beliefs count as the same when they have the same sense (in Perry's usage) or when they consist in relations to internal sentences with the same character. In fleeing propositions, however, we have given up one of their useful features: language neutrality. The demand that we construe all like-believers (in the new, psychologically realistic type-casting) as thinking in the same language is apparently onerous, unless of course some way can be found either to defend this implication or to trivialize it.[8]

How might we defend sententialism? By declaring that it is an empirical question, and an interesting and important one at that, whether people do think in the same or different languages of thought. Perhaps dogs think in Doggish and people think in Peoplish. Perhaps we will discover 'the brain writing which people have in common regardless of their nationality and other differences' (Zeman 1963). Such a discovery would certainly be a theoretical treasure. Or perhaps there will turn out to be non-trivially different Mentalese languages (not mere notational variants of each other) so that the implication we will have to live with is that if your brain speaks Mentallatin while mine speaks Mentalgreek, we cannot indeed *share the same psychological states*. Any comparison would 'lose something in translation'. This would not prevent us, necessarily, from sharing *propositional* attitudes (no longer considered 'pure' psychological states); your way of believing that whales are mammals would just be different—in psychologically non-trivial ways—from mine. It would be a theoretical calamity to discover that each person thinks in a different, entirely idiosyncratic Mentalese, for then psychological generalization would be hard to come by, but if there turned out to be a smallish number of different languages of thought (with a few dialects thrown in) this might prove as theoretically fruitful as the monolingual discovery, since we might be able to explain important differences in cognitive styles by a multilingual hypothesis. (E.g., people who think in Mentallatin do better on certain sorts of reasoning problems, while people who think in Mentalgreek are superior analogy-discoverers. Recall all the old saws about English being the language of commerce, French the language of diplomacy and Italian the language of love.)

In any event, there is a strategy of theory-development available which will tend to replace multilingual hypotheses or interpretations with monolingual hypotheses or interpretations. Suppose we have tentatively fixed a level of functional description of two individuals according to which they are not colingual; their psychological states do not share a syntax. We can cast about for a somewhat higher level of abstraction at which we can redescribe their psychological states so that what had heretofore been treated as syntactic differences were now dismissed as physical differences *beneath syntax*. At the higher functional level we will discover *the same function* being subserved by what we had been considering syntactically different items, and this will entitle us to declare the earlier syntactic taxonomy too fine-grained (for distinguishing what are merely notational variants of rival tokening systems). Availing ourselves of this strategy will tend to blur the lines between syntax and semantics, for what we count as a syntactic feature at one level of analysis will depend on its capacity to figure in semantically relevant differences at the next higher level of functional analysis. Carried to a Pickwickian extreme, we would find ourselves at a *very* abstract level of functional analysis defending a version of monolingualism for Mentalese analogous to the following claim about natural language: French and English are just notational variants of one another or of some *ur*-language; 'bouche' and 'mouth' are different tokens of the same type (cf. Sellars 1974). Normally, and for good reasons, we consider these two words to share only *semantic* properties. Similar principles would presumably constrain our theorizing about Mentalese and its dialects. Anyone who thinks that there *has* to be a single Mentalese for human beings must be ignoring the existence of such principles and falling for the Pickwickian version of monolingualism.

Even if we acquiesced in multilingualism, and found it theoretically fruitful to conduct psychological investigations at such a fine-grained level that we could tolerate psychological state type-casting in syntactical terms, we would still want to have a way of pointing to important psychological *similarities* between counterpart states in people thinking in different languages of thought—analogous to the similarity between '*j'ai faim*' and 'I'm hungry' and '*Ich habe Hunger*', and between '*Es tut mir Leid*' and 'I'm sorry'. If we wish to view the claims we make at this level of abstraction as claims about *tokens of the same type*, the type-casting in question cannot be syntactic (for even at the grossest syntactic level, these expressions are grammatically different, having nouns where others have adjectives, for instance), nor can it be wholly semantic in the sense of *propositional*—since the expressions, with their indexicals, express different propositions on different occasions. What we will need is an *intermediate* taxonomy: the similar items will be similar in that they have similar *roles* to play within a functionalistic theory of the believers (cf. Sellars 1974).[9]

Kaplan is mute on the question of whether his notion of character can be applied interlinguistically. Does '*j'ai faim*' have just the same character as 'I'm hungry'? We will want our psychological counterpart to character to have this feature if we are to use it to characterize psychological similarities that can exist between believers who think in different Mentalese languages. For recall that we are searching for a way of characterizing in the most general way the organismic contribution to the fixation of propositional attitudes, and since we want to grant that you and I can both believe the proposition that whales are mammals in spite of our Mentalese differences, we need to characterize that which is in common in us that can sometimes yield the same function from contexts—embeddings in the world—to propositions.

The value of a syntax-neutral level of pyschological characterization emerges more clearly when one considers the task that faces the sentential attitude theorist who hopes to characterize a human being—more specifically a human nervous system—at a purely syntactical, utterly uninterpreted level of description. (This would provide the raw material, the 'text', for subsequent semantic interpretation.) This would be methodological solipsism or psychology in the narrow sense with a vengeance, for we would be so narrowing our gaze as to lose sight of all the normal relations between things in the environment and the activities within the system. Part of the task would be to distinguish the subset of physical features and regularities inside the organism that betoken syntactic features and regularities—locating and purifying the 'text' midst the welter of scribbles and smudges. Since what makes a feature syntactic is its capacity to make a semantic difference, this purification of text cannot proceed innocent of semantic assumptions, however tentative. How might this work?

Our methodological solipsism dictates that we ignore the environment in which the organism resides—or has resided—but we can still locate a boundary between the organism and its environment, and determine the input and output surfaces of its nervous system. At these peripheries there are the sensory *transducers* and motor *effectors*. The transducers respond to patterns of physical energy impinging on them by producing syntactic objects—'signals'—with certain properties. The effectors at the other end respond to other syntactic objects—'commands' by producing muscle flexions of certain sorts. An idea that in various forms licenses all speculation and theorizing about the semantics of mental representation is the idea that the semantic properties of mental representations are at least partially determinable by their relations, however indirect, with these transducers and effectors. If we know the stimulus conditions of a transducer, for instance, we can *begin* to interpret its signal—subject to many pitfalls and caveats. A similar tentative and partial semantic interpretation of 'commands' can be given once we see what motions of the body they normally produce. Moving towards the centre, downstream from the transducers and upstream from the effectors, we can endow more central events and states with representational powers, and hence at least a partial semantic interpretation. (See, e.g., Dennett 1969, 1978a.)

For the moment, however, we should close our eyes to this information about transducer sensitivity and effector power, and treat the transducers as 'oracles' whose sources of information are hidden (and whose *obiter dicta* are hence uninterpeted by us), and treat the effectors as obedient producers of unknown effects. This might seem to be a bizarre limitation of viewpoint to adopt, but it has its rationale: it is the brain's eye view of the mind, and it is the brain, in the end, that does all the work (see also Dennett 1978b, forthcoming a). Brains are *syntactic engines*, so in the end and in principle the control functions of a human nervous system must be explicable at this level or remain forever mysterious.[10]

The alternative is to hold—most improbably—that *content* or *meaning* or *semantic value* could be independent, detectable causal properties of events in the nervous system. To see what I mean by this, consider a simpler case. There are two coins in my pocket, and one of them (only) *has spent exactly ten minutes on my desk*. This property is not a property causally relevant to how it will affect any entity it subsequently comes in contact with. There is no coin-machine, however sophisticated, that could reject the coin by testing it for *that* property—though it might reject it for being radioactive or greasy or warmer than room temperature. Now if the coin had one of these properties just in virtue of having spent exactly ten minutes on my desk (the desk is radioactive, covered with grease, a combination desk and pottery kiln) the coin-machine could be used to test indirectly (and of course not very reliably) for the property of having spent ten minutes on my desk. The brain's testing of *semantic* properties of signals and states in the nervous system must be similarly indirect testing, driven by merely syntactic properties of the items being discriminated—that is, by whatever structural properties the items have that are amenable to direct mechanical test.[11] Somehow, the syntactical virtuosity of our brains permits us to be interpreted at another level as *semantic engines*—systems that (indirectly) discriminate the significance of the impingements on them, that understand, mean and believe.

This vantage point on brains as syntactic engines gives us a diagnosis of what is going on in Putnam's, Kaplan's and Perry's arguments. If the meaning of anything—e.g., an internal information-storing state, a perceived change in the environment, a heard utterance—is a property that is only indirectly detectable by a system such as a person's brain, then meaning so conceived is *not* the property to use to build projectible predicates descriptive of the system's behaviour. What we want is rather a property that is to meaning so conceived roughly as the property *guilt-beyond-a-reasonable-doubt* is to *guilt*. If you want to predict whether the jury will acquit or convict, the latter property is unfortunately but unavoidably a bit less reliable than the former.

But then we can see that Putnam *et al.* are driving an ancient wedge into a new crack—the distinction between *real* and *apparent* is being tailored to distinguish real and apparent propositions believed. Then it is not surprising, but also not very cheering, to note that the theoretical move many want to make in this situation is analogous to the move that earlier landed philosophers with sense-data, or with qualia. What is *directly* accessible to the mind is not a feature of the surfaces of things *out there*, but a sort of inner copy that has a life of its own. Sentences of Mentalese are seen from this vantage point to be inner copies of the propositions we come to believe in virtue of our placement in the world. One must not argue guilt by association, however, so it remains an open question whether in this instance the theoretical move can yield us a useful model of the mind—whatever its shortcomings in earlier applications. We should persist with our earlier question: what could we understand about a brain (or a mind) considered just as a syntactic engine?

The strategy of methodological solipsism mates with the language of thought model of mentality to produce the tempting idea that one could in principle divide psychology into syntactic psychology (pursued under methodological solipsism) and semantic psychology (which would require one to cast one's glance out at the world). We have seen that the preliminary task of discovering which internal features ought to be considered syntactical depends on assumptions about the semantic roles to be played by events in the system, but it is tempting to suppose that the syntax of the system will not depend on particular details of these semantic roles, but just on assumptions about the existence and differentiation of these roles. Suppose, runs this tempting line of thought, we honour our methodological solipsism by *de-interpreting* the messages sent by the transducers, and the commands sent to the effectors: transducers are then taken to assert only that it is F now, getting G-er and G-er, intermittently H—where these are uninterpreted sensory predicates; and effectors obediently turn on the X-er or the Y-er, or cause the Z to move. Might we not then be able to determine the *relative* semantic interpretation of more central states (presumably beliefs, desires, and such) in terms of these uninterpreted predicates? We would be able to learn that the system's past history had in one way or another brought it into the state of believing that all Fs are very GHs, and that X-ing usually leads to either a JK or a JL. The idea that we might do this is parallel to Field's suggestion (1972) that we might do Tarskian semantics for a (natural) language in two independently completable parts: the theory of reference for the primitives, and everything else. We can do the latter part first while temporizing by the use of such statements of primitive reference as

'snow' refers to whatever it refers to

The enabling idea of sentential attitude psychology is that we might similarly be able to temporize about the ultimate reference of the predicates of Mentalese, while proceeding apace with their relative interpretation, as revealed in a systematic—and entirely internal—semantic edifice. Field (1978) proposes this division of the problem with the help of a technical term, 'believe*'.

(1) X believes that p if and only if there is a sentence S such that X believes* S and S means that p

. . . the effect of adopting (1) is to divide the problem of giving a materialistically adequate account of the belief relation into two sub-problems:

subproblem (a): the problem of explaining what it is for a person to believe* a sentence (of his or her own language).

subproblem (b): the problem of explaining what it is for a sentence to mean that *p*.

... The rough idea of how to give an account of (a) should be clear enough: I believe* a sentence of my language if and only if I am disposed to employ that sentence in a certain way in reasoning, deliberating and so on. This is very vague of course ... but I hope that even the vague remarks above are enough to predispose the reader to think that believing* is not a relation that should be a particular worry to a materialist (even a materialist impressed by Brentano's problem [of intentionality]). On the other hand, anyone impressed with Brentano's problem *is* likely to be impressed with subproblem (b), for unlike (a), (b) invokes a *semantic* relation (of *meaning that*). [1978, p. 13.]

So believing* is supposed to be an entirely *non*-semantic relation between a person and a syntactically characterized object. The 'certain way' one must be disposed to employ the sentence is left unspecified, of course, but the presumption must be that its specification can in principle be completed in syntactic terms alone. Only thus could the relation not be a 'worry' to a materialist.

So long as Field stays with sentences of natural language as the *relata* of *believes**, he is on safe enough ground speaking in this way, for a sentence of a natural language can be identified independently of any person's dispositions to use it in various ways. But once Field turns to Mentalese sentences (or sentence-analogues, as he calls them)—as he must, for familiar reasons having to do with mute, animal and prelinguistic believers, for instance—this first-approximation definition of 'believe*' becomes highly problematic—although Field himself characterizes the shift as 'only a minor modification' (p. 18).

Take just the simplest case: the 'message' sent by a relatively peripheral reporter element near the retina. Let us call this element Rep. Let us suppose our first hypothesis is that Rep's signal is a token of the Mentalese sentence (in strict English translation) 'There is now a small red spot in the middle of the visual field.' De-interpreting the sentence, we see it to have the syntactical shape (for our purposes—and what other purposes could matter?) *there is now an FGH at J of K.* We take there to be this many terms in the message just because we suppose the message can contribute in this many different ways in virtue of its links. But perhaps we've misinterpreted its function in the system. Perhaps the Mentalese sentence to associate with it (again in strict English translation) is 'There's a tomato in front of me' or merely 'At least ten retinal cells of sort F are in state G' or perhaps 'I am being appeared to redly'. These sentences (at least their English translations as shown) have quite different syntactical analyses. But which syntactical form does the thing we have located in the brain have? We may be able to determine the 'shape' of an item—an event-type, for instance—in the brain but we can't determine its syntactical form (as distinguished from its merely decorative—however distinctive—properties) except by determining its particular powers of combination and cooperation with the other elements, and ultimately its *environmental* import via those powers of interaction.

A thought-experiment will bring out the point more clearly. Suppose that our task were *designing* a language of thought rather than discovering an already existent language of thought in operation. We figure out what we want our system to believe (desire, etc.) and write down versions of all this information in sentences of some tentative variety of Mentalese. We inscribe each belief-sentence in a separate box of a large map of the system we are designing. One belief box has the Mentalese translation of 'snow is white' in it. Now just having the symbols written in the box cannot store the information that snow is white, of course. At the very least there must be machinery poised to utilize these symbols in this box in a way that makes a difference—the right sort of difference for believing that snow is white. This machinery must, for instance, somehow link the box with 'snow is white' in it to all the boxes in which sentences with the Mentalese word for 'white' occurs. These boxes are linked to each other, in some systematic ways, and ultimately somehow to the periphery of the system—to the machinery that could signal the presence of cold white stuff, cold green stuff, and so forth. The 'snow is white' box is also linked to all the 'snow' boxes, which are linked to all the 'precipitation' boxes and so forth. The imagined vast network of links would turn the collection of boxes into *something like* the taxonomic lattices or structural inheritance networks or semantic nets to be found in Artificial Intelligence systems (see, e.g., Woods 1975, forthcoming). Then there are all the links to the machinery, whatever it is, that relies on these boxes to contribute appositely to the control of the behaviour of the whole system—or better: the creature in which the system is embedded. Without all these links, the inscriptions in the boxes are mere decoration—they don't store the information whatever they look like. But equally, once the links are in place, the inscriptions in the boxes are still mere decoration—or at best mnemonic labels encapsulating *for us* (more or less accurately) the information actually stored *for the system* at that node in virtue of the links from the node to other nodes of the system. The real 'syntax', the structure in the system on which function depends, is all in the links.[12]

The separation imagined between the links and the inscriptions in the boxes in our example does not of course reflect the actual situation in Artificial Intelligence. The point of computer languages is that they are cleverly designed so that their inscriptions, properly entered in the system, *create* various links to elements in other inscriptions. It is this feature that makes computer languages so different from natural languages, and surely it is a feature that any language of thought would have to have—if its postulation is to avoid the fruitless epiphenomenalism of our imagined labels in boxes. And for any 'language' having this feature, the relation between form and function is tight indeed—so tight that the distinction in Kaplan's enlarged schema at stage (0) between the contribution of syntactic features and the contribution of 'linguistic conventions' has no counterpart in any plausible 'language of thought' model for psychology. Getting the 'text' independently of getting its interpretation is not a real prospect for psychology. So Field's proposed division of the problem of belief into subproblem (a), the syntactic problem, and subproblem (b), the semantic problem, if taken as a proposal for a research strategy, is forlorn.

Perhaps, however, Field's proposal can be recast, not as a

recommendation for a research programme, but as marking an important distinction of reason. Even if we cannot in fact *first* determine the syntax and *then* the semantics of a language of thought (for epistemological reasons), the distinction might still mark something real, so that, having bootstrapped our way to both a semantic and a syntactic theory of the language of thought, we could distinguish the system's syntactic properties from its semantic properties. No doubt a distinction rather like the distinction between semantics and syntax will be makeable in retrospect within any mature and confirmed psychology of belief, for there must be some way of describing the operation of the nervous system independently of its embedding in the world in virtue of which we fix its semantic characterization. But supposing that this distinction will have much in common with the distinction between syntax and semantics for a natural language is committing oneself to a gratuitously strong sententialism.

For a stronger moral can be drawn from the discussion of the problem of associating an *explicit message* with the contribution of Rep, the peripheral visual transducer. We supposed that we had isolated Rep as a functional component of the cognitive system, a component that could be seen to *inform* the system about some visual feature. Then we faced the problem of coming up with an apt *linguification* of that contribution—finding, that is, a *sentence* that explicitly and accurately expresses the message being asserted by Rep. And we saw that which sentence we chose depended critically on just what combinatorial powers Rep's message actually had. Even supposing we could determine this, even supposing that we could establish that we had a *best* functional description of the system of which Rep is a part, and could say *exactly* what functions Rep's signal can perform, there is no guarantee that *those* functions will be aptly and accurately referred to or alluded to by *any* claim to the effect that Rep's message inserts sentence S (in language L) as a premise in some deductive or inferential system. The conviction that it *must* be possible to linguify any such contribution seems to me to lie beneath a great deal of recent meta-theoretical ideology, and I suspect it rests in part on a mistaken conflation of *determinateness* and *explicitness*.[13] Suppose Rep's semantic contribution, its *informing* of the system, is entirely determinate. That is, we can say exactly how its occurrence would produce effects that ramify through the system making differences to the content or semantic contribution of other subsystems. Still it would not follow that we could render this contribution *explicit* in the form of a sentence or sentences asserted. We might have some way of *describing* the semantic contribution perfectly explicitly, without describing it as any explicit assertion in any language.

I am not merely alluding to the possibility that activity in one part of a cognitive system might have a systematic but noisy—or at any rate non-contentful—effect on another part. Such effects are quite possible. Smelling sulphur might just make someone think about baseball *for no reason*. That is, there could be no *informational* link, such as memories of odorous games behind the gasworks, but nevertheless a reliable if pointless *causal* link. I grant the possibility of such effects, but am urging something stronger: that there could be highly content-sensitive, informational, epistemically useful, designed relations between activities in different cognitive subsystems which nevertheless defied sententialist interpretation. Suppose, for instance, that Pat says that Mike 'has a thing about redheads'. What Pat means is, roughly, that Mike has a stereotype of a redhead which is rather derogatory, and which influences Mike's expectations about and interactions with redheads. It's not just that he's prejudiced against redheads, but that he has a rather idiosyncratic and *particular* thing about redheads. And Pat might be right—more right than he knew! It could turn out that Mike does have this thing, this bit of cognitive machinery, that is *about redheads* in the sense that it systematically comes into play whenever the topic is redheads or a redhead, and which adjusts various parameters of the cognitive machinery, making flattering hypotheses about redheads less likely to be entertained, or confirmed, making relatively aggressive behaviour *vis-à-vis* redheads closer to implementation than otherwise it would be, and so forth. Such a *thing about redheads* could be very complex in its operation, or quite simple, and in either case its role could elude characterization in the format:

Mike believes that: $(x)(x$ is a redhead $\supset \ldots)$

no matter how deviously we piled on the exclusion clauses, qualifiers, probability operators, and other explicit adjusters of content. The contribution of Mike's thing about redheads could be perfectly determinate and also undeniably contentful and yet no linguification of it could be more than a mnemonic label for its role. In such a case we could say, as there is often reason to do, that various beliefs are *implicit* in the system. For instance, the belief that redheads are untrustworthy. Or should it be the belief that most redheads are untrustworthy; or 'all the redheads I have met'? Or should it be '$(x)(x$ is a redhead $\supset$ the probability is 0.9 that x is untrustworthy)'? The concern with the proper form of the sentence is idle when the sentence is only part of a stab at capturing the implicit content of some non-sentential bit of machinery. (Cf. section II above on the futile debates occasioned by failure to distinguish propositions from sentences.)

It still might be argued that although *non*-sentential semantic contributors of the sort I have just sketched could play a large role in the cognitive machinery of the human brain, explicitly sentential representational states are also required, if only to give the finite brain the compositionality it needs to represent indefinitely many different states of affairs with its finite resources. And for all I can see this may be so. Certainly some very efficient and elegant sort of compositionality accounts for the essentially limitless powers we have to perceive, think about, believe, intend, . . . different things. The only examples we now have of (arguably) *universal* systems of representation with finite means are languages, and perhaps any possible universal system of representation must be recognizably sentential, in a sense yet to be settled, of course. Suppose this is so (and find yourself in good company). Then although a great deal of psychology might be both cognitive and *non*-sentential, at the core of the person would be his sentential system. The theory of visual perception, for instance, might require Mentalese only at some relatively central 'interface' with the sentential core. Initially we viewed the task of sentential attitude psychology as starting at the peripheral transducers and effectors and sententializing all the way in. Perhaps the mistake was just in supposing that the interface between the sentential system *and the world*—the overcoat of transducers and effectors worn by every cognitive system—was thinner than it is.

Thick or thin, the overcoat of transducer–perceiver mechanisms and effector–actor mechanisms becomes an environment of sorts, a context, in which the postulated 'utterances' of Mentalese occur. Ignoring that context, the predicates of Mentalese are *very* uninterpreted; even taking that context into account, the predicates of Mentalese would be only partially interpreted—not fully enough interpreted, for instance, to distinguish my propositional attitudes from my *Doppelgänger's*. We can construct some bizarre variations on Putnam's thought-experiment using this notion. Suppose a physical duplicate of that part of my nervous system which is the sentential system were hooked up (right here on Earth) to a different overcoat of transducers and effectors. My sentential system can store the information that all *F*s are very *GH*s, and so can its replica, but in me, this sentential state subserves[14] my belief that greyhounds are very swift animals, while in the other fellow it subserves the belief that palaces are very expensive houses! These different tokens of the same *syntactic* type have the same *character* if we treat the transducer–effector overcoat as an undifferentiated part of the 'external' context of utterance; if we draw a further boundary between the overcoat and the environment, then these tokens do not have the same character, but only the same syntax, and the different overcoats play the counterpart roles of the linguistic conventions of Kaplan's stage (o).

The point is that Kaplan's schema is a special case of something very general. Whenever we are describing a functional system, if we draw a boundary between the system 'proper' and some context or environmental niche in which it resides, we find we can characterize a Kaplan-style schema

$$C + E \rightarrow I$$

where C is a character-like concept of narrow or intra-systemic application; E is the concept of an embedding context or environment of operation, and I is a *richer* semantic (or functional) characterization of the systemic role in question than that provided by C alone. Where the system in question is a representing or believing system, 'richer' means closer to determining a (classical) proposition, or, if we include Kaplan's stage (2) as the ultimate step in this progression, richer in the sense of being closer to ultimate reference to things in the world. In other contexts—such as characterizations of functional components in biology or engineering (see Wimsatt 1974)—the 'richer' characterization tells us more about the functional point of the item: what is narrowly seen as a spark-producer is seen, in context, to be a fuel-igniter, to take an overworked example.

Moving from stage to stage in such an interpretation schema, one sees that the richer the semantics of a particular stage, the more abstract or tolerant the syntax. Sentences with *different physical properties* can have the *same syntax*. Sentences with *different syntax* can have the *same character*. Sentences with *different character* can express the *same proposition*. *Different propositions*, finally, can attribute the same property to the same individual: that the Dean of Admissions is middle-aged is not identical with the proposition that the tallest Dean is middle-aged. Transplanted from the theory of natural language to the theory of psychological states, the part of the nesting of concern to us now looks like this: people believing the same proposition can be in different (narrow) psychological states; people in the same narrow psychological state can be in different fine-grained (i.e., syntactically characterized) states; people in the same syntactic states can implement those states in physically different ways. And, of course, looking in the other direction, we can see that two people narrowly construed as being in the same state can be reconstrued as being in different states if we redraw the boundaries between the people's states and the surrounding environment.[15]

IV. Notional Attitudes

In the face of the objections of Putnam and others to 'classical' propositional attitudes, we adverted to the question: what is the *organismic contribution* to the fixation of propositional attitudes? The answer would characterize psychological states 'in the narrow sense'. The attempt to capture these narrow psychological state types as *sentential attitudes* ran into a variety of problems, chief of which was that any sentential attitude characterization, being essentially a *syntactical* type-casting, would cut too fine. In Putnam's thought-experiment we grant that *physical* replicahood is sufficient but not necessary for identity of organismic contribution; we could also grant that the weaker similarity captured by *syntactic* replicahood (at some level of abstraction) would be sufficient for identity of organismic contribution, but even though identity of organismic contribution—narrow-psychological twinhood—is a very stringent condition, it would not seem to *require* syntactic twinhood, at any level of description. Consider the somewhat analogous question: do all Turing machines that compute the same function share a syntactic (i.e., machine table) description? No, unless we adjust our levels of description of the machine table and the input–output behaviour so that they coalesce trivially. What should *count* as equivalence for Turing machines (or computer programs) is a vexed question; it would not be if it weren't for the fact that non-trivially different descriptions in terms of internal 'syntax' can yield the *same* 'contribution'—at some useful level of description.

The analogy is imperfect, no doubt, and other considerations— e.g., biological considerations—might weigh in favour of supposing that *complete* narrow-psychological twinhood required syntatic twinhood at some level, but even if that were granted, it would not at all follow that *partial* psychological similarity can always be described in some general system of syntactic description applicable to all who share the psychological trait. People who are vain, or paranoid, for instance, are surely psychologically similar; a large part of the similarity in each case would seem well captured by talking of similar or shared beliefs. Even if one takes a self-defeatingly stringent line on belief-identity (according to which no two people ever *really* share a belief), these *similarities* in belief cry out for capturing within psychology. They could not plausibly be held to depend on monolingualism—vain people's brains all speaking the same Mentalese. Nor can we capture these similarities in belief-state via *propositional* attitudes, because of the indexicality of many of the crucial beliefs: 'People admire *me*', 'People are trying to ruin *me*'.

These considerations suggest that what we are looking to characterize is an intermediate position—halfway between syntax and semantics you might say. Let us call it *notional attitude psychology*. We want it to work out that I and my *Doppelgänger*— and any other narrow-psychological twins—have exactly the same notional attitudes, so that our differences in propositional attitudes are due entirely to the different environmental contributions. But we also want it to work out that you and I, no psychological twins but 'of like mind' on several topics, share a variety of notional attitudes.

A familiar idea that has occurred in many guises can be adapted for our purposes here: the idea of a person's subjective world, Helen Keller's *The World I Live In*, or John Irving's *The World According to Garp*, for instance. Let us try to characterize the *notional world* of a psychological subject so that, for instance, although my *Doppelgänger* and I live in different real worlds— Twin-Earth and Earth—we have the *same* notional world. You and I live in the same real world, but have different notional worlds, though there is a considerable overlap between them.

A notional world should be viewed as a sort of *fictional* world devised by a theorist, a third-party observer, in order to characterize the narrow-psychological states of a subject. A notional world can be supposed to be full of notional objects, and the scene of notional events—all the objects and events the subject *believes in*, you might say. If we relax our methodological solipsism for a moment, we will note that some objects in the real world inhabited by a subject 'match' objects in the subject's notional world, but others do not. The real world contains many things and events having no counterparts in any subject's notional world (excluding the notional world of an omniscient God), and the notional worlds of gullible or confused or ontologically profligate subjects will contain notional objects having no counterparts in the real world. The task of describing the relations that may exist between things in the real world and things in someone's notional world is notoriously puzzle-ridden—

that is one reason to retreat to methodological solipsism: to factor out those troublesome issues temporarily.

Our retreat has landed us in very familiar territory: what are notional objects but the *intentional objects* of Brentano? Methodological solipsism is apparently a version of Husserl's époché, or bracketing. Can it be that the alternative to both propositional attitude psychology and sentential attitude psychology is . . . Phenomenology? Not quite. There is one major difference between the approach to be sketched here and the traditional approaches associated with Phenomenology. Whereas Phenomenologists propose that one can get to one's *own* notional world by some special somewhat introspectionist bit of mental gymnastics—called, by some, the phenomenological reduction—we are concerned with determining the notional world of *another*, from the outside. The tradition of Brentano and Husserl is *auto-phenomenology*; I am proposing *hetero-phenomenology*.[16] Although the results might bear a striking resemblance, the enabling assumptions are very different.

The difference can best be seen with the aid of a distinction recently resurrected by Fodor (1980) between what he calls, following James, *naturalistic* and *rational* psychology. Fodor quotes James:

> On the whole, few recent formulas have done more service of a rough sort in psychology than the Spencerian one that the essence of mental life and of bodily life are one, namely, 'the adjustment of inner to outer relations'. Such a formula is vagueness incarnate; but because it takes into account the fact that minds inhabit environments which act on them and on which they in turn react; because, in short, it takes mind in the midst of all its concrete relations, it is immensely more fertile than the old-fashioned 'rational psychology' which treated the soul as a detached existent, sufficient unto itself, and assumed to consider only its nature and its properties. [James 1890, p. 6.]

James sings the praises of naturalistic psychology, psychology in the *wide* sense, but the moral from Twin-Earth, drawn explicitly by Fodor, is that naturalistic psychology casts its net too wide to be do-able. The Phenomenologists draw the same conclusion, apparently, and both turn to different versions of methodological solipsism: concern for the psychological subject 'as a detached existent, sufficient unto itself', but when they 'consider its nature and its properties', what do they find? The Phenomenologists, using some sort of introspection, claim to find a *given* in experience, which becomes the raw material for their construction of their notional worlds. If Fodor, using some sort of (imagined) internal inspection of the machinery, claimed to find a Mentalese text *given* in the hardware (which would become the raw material for construction of the notional attitudes of the subject) we would have as much reason to doubt the existence of the given in this case as in the case of Phenomenology. James is right: you cannot do *psychology* (as opposed to, say, neurophysiology) without determining the *semantic* properties of the internal events and structures under examination, and you cannot uncover the semantic properties without looking at the relations of those internal events or structures to things in the subject's environment. But nowhere is it written that the environment relative to which we fix such a system's semantic properties must be a *real* environment, or the *actual* environment in which the system has grown up. A fictional environment, an idealized or imaginary environment, might do as well. The idea is that in order to do 'mental representation' theory, you need to do the semantics of the representations *from the beginning*. (You can't first do the syntax, then the semantics.) But that means you need a model, in the sense of Tarskian semantics. A fictional model, however, might permit enough Tarskian semantics to get under way for us to determine the partial semantics, or proto-semantics, we need to characterize the organismic contribution.

The idea of a notional world, then, is the idea of a model—but not necessarily the actual, real, true model—of one's internal representations. *It does not consist itself of representations but of representeds.* It is the world 'I live in', not the world of representations *in me*. (So far, this is pure Brentano, at least as I understand him. See Aquila 1977.) The theorist wishing to characterize the narrow-psychological states of a creature, or in other words, the organismic contribution of that creature to its propositional attitudes, *describes* a fictional world; the description exists on paper, the fictional world does not exist, but the inhabitants of the fictional world are treated as the notional referents of the subject's representations, as the intentional objects of that subject. It is hoped that by this ploy the theorist can get the benefits of James's and Spencer's naturalism without the difficulties raised by Putnam and the rest.

Now the question is: what guides our construction of an organism's notional world? Suppose, to dramatize the problem, we receive a box containing an organism from we know not where, alive but frozen (or comatose)—and hence cut off from any environment. We have a Laplacean snapshot of the organism—a complete description of its internal structure and composition—and we can suppose that this enables us to determine exactly how it *would* respond to any new environmental impacts were we to release it from its state of suspended animation and isolation. Our task is like the problem posed when we are shown some novel or antique gadget, and asked: what is it for? Is it a needle-making machine or a device for measuring the height of distant objects, or a weapon? What can we learn from studying the object? We can determine how the parts mesh, what happens under various conditions, and so forth. We can also look for tell-tale scars and dents, wear and tear. Once we have compiled these facts *we try to imagine a setting* in which given these facts it would *excellently* perform some imaginably useful function. If the object would be an equally good sail-mender or cherry-pitter we won't be able to tell what it *really* is—what it is *for*—without learning where it came from, who made it, and why. Those facts could have vanished without a trace. Such an object's true identity, or essence, could then be utterly undeterminable by us, no matter how assiduously we studied the object. That would not mean that there was no fact of the matter about whether the thing was a cherry-pitter or a sail-mender, but that the truth, whichever it was, no longer made a difference. It would be one of those idle or inert historical facts, like the fact that some of the gold in my teeth once belonged to Julius Caesar—or didn't.

Faced with our novel organism, we can easily enough determine what it is for—it is for surviving and flourishing and reproducing its kind—and we should have little trouble identifying its sense organs and modes of action and biological needs. Since *ex hypothesi* we can figure out what it would do if ..., we can determine, for instance, that it will eat apples, but not fish; tends to avoid brightly lit places, is disposed to make certain noises under certain conditions, and ... Now what kind of an environment would these talents and proclivities fit it for? The more we learn about the internal structure, behavioural dispositions and systemic needs of the organism, the more particular becomes our hypothetical ideal environment. By 'ideal environment' I do not mean the best of all possible worlds for this organism ('... with the lemonade spring where the bluebirds sing ...'), but the environment (or class of environments) for which the organism as currently constituted is best fitted. It might be a downright nasty world, but at least the organism is prepared to cope with its nastiness. We can learn something about the organism's enemies—real or only notional—by noting its protective coloration, or its escape behaviour, or ... how it would answer certain questions.

So long as the organism we are dealing with is very simple, and has, for instance, little or no plasticity in its nervous system (so it cannot learn), the limit of specificity for the imagined ideal environment may fail to distinguish radically different but equally well-fitted environments, as in the case of the gadget. As the capacity to learn and remember grows, and as the richness and complexity of the possible relations with environmental conditions grows (see Dennett forthcoming b), the class of equally acceptable models (hypothetical ideal environments) shrinks. Moreover, in creatures with the capacity to learn, and store information about their world in memory, a new and more powerful exegetical principle comes into play. The scars and dents on the cherry-pitter (or was it a sail-mender?) may on occasion prove to be tell-tale, but the scars and dents on a learning creature's memory are *designed* to be tell-tale, to record with high fidelity both particular encounters and general lessons, for future use. Since the scars and dents of memory are for future use, we can hope to 'read' them by exploiting our knowledge of the dispositions that depend on them, so long as we assume the dispositions so attached are in general appropriate. Such interpretations of 'memory traces' yield more specific information about the world in which the creature lived and to which it had accommodated itself. But we will not be able to tell information about this world from misinformation, and thus the world we extrapolate as *constituted* by the organism's current state will be an ideal world, not in the sense of *best*, but in the sense of *unreal*.

The naturalists will rightly insist that the actual environment as encountered has left its mark on the organism and intricately shaped it; the organism is in its current state *because of the history it has had*, and only such a history could in fact have put it into its present state. But in a thought-experimental mood we can imagine creating a duplicate whose *apparent* history was not its actual history (as in the case of a faked antique, with its simulated 'distress' marks and wear and tear). Such a complete duplication (which is only logically, thought-experimentally, possible) is the limiting case of something actual and familiar: any particular feature of current state may be misbegotten, so that the way the world ought to have been for the creature now to be in this state is not quite the way the world was. The notional world we describe by extrapolation from current state is thus not exactly the world we take to have created that state, even if we know that actual world, but rather the apparent world of the creature, the world apparent *to* the creature as manifested in the creature's current total dispositional state.

Suppose we apply this imaginary exercise in notional world formation to highly adaptive organisms like ourselves. Such organisms have internal structure and dispositional traits so rich in information about the environment in which they grew up that we could in principle say: this organism is best fitted to an environment in which there is a city called Boston, in which the organism spent its youth, in the company of organisms named . . . and so forth. We would not be able to distinguish Boston from Twin-Earth Boston, of course, but except for such virtually indistinguishable variations on a theme, our exercise in notional world formation would end in a unique solution.

That, at any rate, is the myth. It is a practically useless myth, of course, but theoretically important, for it reveals the fundamental assumptions that are being made about the ultimate dependence of the organismic contribution on the physical constitution of the organism. (This dependence is otherwise known as the supervenience of (narrow) psychological traits on physical traits; see, e.g., Stich 1978b.) At the same time, the myth preserves the *under*determination of ultimate reference that was the acclaimed moral of the Putnamian considerations. If there is a language of thought, this is how you would have to bootstrap your way to discovering it and translating it—without so much as the benefit of bilingual interpreters or circumstantial evidence about the source of the text. If there is any third-person alternative to the dubious introspectionist (*genuinely* solipsistic) method of the Phenomenologists, if hetero-phenomenology is possible at all, it will have to be by this method.

In principle, then, the ultimate fruits of the method, applied to a human being under the constraints of methodological solipsism, would be an exhaustive description of that person's notional world, complete with its mistaken identities, chimaeras and personal bogeymen, factual errors and distortions.[17] We may think of it as *the* notional world of the individual, but of course the most exhaustive description possible would fail to specify a unique world. For instance, variations in a world entirely beyond the ken, or interests, of a person would generate different possible worlds equally consistent with the maximal determination provided by the constitution of the person.

The situation is analogous to that of more familiar fictional worlds, such as the world of Sherlock Holmes, or Dickens's London. Lewis (1978) provides an account of 'truth in fiction', the semantics of the interpretation of fiction, that develops the idea we need: 'the' world of Sherlock Holmes is formally better conceived of as a *set* of possible worlds, roughly: all the possible worlds consistent with the entire corpus of Sherlock Holmes texts by Conan Doyle.[18] Similarly, 'the' notional world we describe might better be viewed formally as the set of possible worlds consistent with the maximal description.[19] Note that the description is the *theorist's* description; we do not *assume* that the structural features of the organism on which the theorist bases his description include elements which themselves are descriptions. (The features of the cherry-pitter that lead us to describe a cherry (rather than a peach or an olive) are not themselves descriptions of cherries.) From this perspective, we can see that Putnam has devised Twin-Earth and Earth both to be members of the set of possible worlds that *is* the notional world I share with my *Doppelgänger*. XYZ slakes thirst, dissolves wallpaper paste and produces rainbows as well as H_2O does; its difference from H_2O is beneath all the thresholds of discrimination of both me and my *Doppelgänger*—provided, presumably, that neither of us is, or consults with, a wily chemist or microphysicist.

Given a notional world for a subject, we can talk about what the subject's beliefs are *about*, in a peculiar, but familiar, sense of 'about'. Goodman (1961) discusses sentences of Dickens that are 'Pickwick-about', a semantic feature of these sentences that is not genuinely relational, there being no Mr. Pickwick for them to be about in the strong, relational sense. In a similar spirit, Brentano discusses the 'relation*like*' status of mental phenomena whose intentional objects are non-existent. (See Aquila 1977.) An enabling assumption of notional attitude psychology is that the theorist can use Pickwick-aboutness and its kin as the semantic properties one needs for the foundations of any theory of mental representation.

The strategy is not untried. Although notional attitude psychology has been concocted here as a response to the philosophical problems encountered in propositional attitude and sentential attitude psychology, it can be easily discerned to be the tacit methodology and ideology of a major branch of Artificial Intelligence. Consider, for instance, the now famous SHRDLU system of Winograd (1972). SHRDLU is a 'robot' that 'lives in' a world consisting of a table on which there are blocks of various colours and shapes. It perceives these and moves them about in response to typed commands (in English) and can answer questions (in English) about its activities and the state of its world. The scare-quotes above are crucial, for SHRDLU isn't really a robot, and there is no table with blocks on it for SHRDLU to manipulate. That world, and SHRDLU's actions in it, are merely simulated in the computer program of which SHRDLU the robot *simulation* is a part. Fodor (1980) makes the point we want, even anticipating our terminology:

In effect, the machine lives in an entirely notional world; all its beliefs are false. Of course, it doesn't matter to the machine that its beliefs are false since falsity is a semantic property and, qua computer, the device satisfies the formality conditions; *viz.*, it has access only to formal (non-semantic) properties of the representation that it manipulates. In effect, the device is in precisely the situation that Descartes dreads: it's a mere computer which dreams that it's a robot. [1980.]

To some critics, the fact that SHRDLU does not really perceive things in the world, touch them, and otherwise enter into causal relations with them, suffices to show that whatever else SHRDLU may have, SHRDLU certainly has no *beliefs* at all. What beliefs could SHRDLU have? What could their content be? What could they be about? SHRDLU is a purely formal system quite unattached to the world by bonds of perception, action or indeed interest. The idea that such merely formal, merely syntactical states and processes, utterly lacking all semantic properties, could provide us with a model of belief is preposterous! (SHRDLU brings out the bluster in people.)

The gentle reply available in principle runs as follows. Indeed, as the critics proclaim, a genuine believer must be richly and intimately attached by perception and action to things in the world, the objects of his belief, but providing those bonds for SHRDLU by providing it with real TV-camera eyes, a real robotic arm, and a real table of blocks on which to live would have been expensive, time consuming, and *of little psychological interest.* Clothed in a transducer–effector overcoat of robotics hardware, SHRDLU would have a notional world of blocks on a table top—which is to say, plunked into that *real* environment SHRDLU would make out very well; a blocks world is a good *niche* for SHRDLU. Stripped of the robotics overcoat, SHRDLU has a vastly less specific notional world; many more possible worlds are indistinguishable to it, but still, the functional structure of the core is the locus of the interesting psychological problems and their proposed solutions, so choosing the blocks world as an *admissible* notional world (it is in the set of Tarski models for the core system) is an innocent way of clothing the system in a bit of verisimilitude.

In the actual case of SHRDLU, this defence would be optimistic; SHRDLU is not that good. It would *not* be trivial, or even an expensive but straightforward bit of engineering, to clothe SHRDLU in robotics, and the reasons why it would not are of psychological interest. By keeping SHRDLU's world merely notional Winograd neatly excused himself from providing solutions to a wealth of difficult, deep, and important problems in psychology. It is far from clear that any improvements on SHRDLU, conceived in the same spirit within the same research programme, could more justly avail themselves of this line of defence. But that it is the assumed ideal line of defence of such research is, I think, unquestionable. To Husserl's claim that bracketing the real world leaves you with the essence of the mental, Winograd and A.I. can add: Yes, and besides, bracketing saves you time and money.

The Husserlian themes in this A.I. research programme are unmistakable, but it is important to remind ourselves as well of the differences. To the auto-phenomenologist, the relative inaccessibility of the real referents of one's beliefs—and hence, as Putnam argues, the relative inaccessibility of one's *propositional* attitudes—is presented as a point about the limits of introspective privileged access, a very *Cartesian* result: *I* cannot discriminate *for sure*; *I* am not *authoritative* about which proposition I am now entertaining, which real object I am now thinking about. But SHRDLU's 'introspections' play no privileged role in Winograd's hetero-phenomenology: SHRDLU's notional world is fixed from the outside by appeal to objective and publicly accessible facts about the capacities and dispositions of the system, and hence its fate in various imagined environments. The counterpart to the Cartesian claim is that even the totality of these *public* facts underdetermines propositional attitudes. Even though the environments appealed to are imaginary, the appeal places hetero-phenomenology squarely on the naturalistic side of the Jamesian division.

The elaboration of imaginary ideal environments for purposes of comparison of internally different systems is a strategy of some currency, in engineering for instance. We can compare the power of different automobile engines by imagining them in pulling contests against a certain fictional horse, or we can compare their fuel efficiency by seeing how far they will push a car in a certain simulated environment. The use of an ideal environment permits one to describe *functional* similarities or competences independently of details of implementation or performance. Utilizing the strategy in psychology to elaborate notional worlds is just a particularly complex case. It enables us to describe partial similarities in the psychological 'competences' of different subjects—for instance, their representational *powers*—in ways that are neutral about their implementation—for instance, their representational *means*.

The analogy with fiction is again useful in making this point. What exactly is the similarity between Shakespeare's *Romeo and Juliet* and Bernstein's *West Side Story?* The latter was 'based on' the former, we know, but what do they actually have in common? Are they about the same people? No, for they are both fictions. Do they contain the same or similar representations? What could this mean? The same words or sentences or descriptions? The scripts of both happen to be written in English, but this is clearly irrelevant, for the similarity we are after survives translation into other languages, and—more dramatically—is evident in the *film* of *West Side Story* and in Gounod's opera. The similarity is independent of any particular *means* of representation—scripts, sketches, descriptions, actors on stages or before cameras—and concerns *what is represented*. It is not any kind of *syntactic* similarity. Since such similarities are as evident in fiction as in factual reporting, we must understand 'what is represented' to take us to elements of a notional world, not necessarily the real world. We can compare different notional or fictional worlds with regard to matters large and small, just as we can compare different parts of the real world. We can compare a notional world with the real world. (The near-sighted Mr. Magoo's notional world only intermittently and partially resembles the real world, but just enough, miraculously, to save him from disaster.)

When, then, shall we say that two different people share a notional attitude or set of notional attitudes? When their notional worlds have a point or region of similarity. Notional worlds are agent-centred or egocentric (Perry 1977; Lewis 1979); when comparing notional worlds for psychological similarity it will typically be useful, therefore, to 'superimpose' the centres—so that the origins, the intersection of the axes, coincide—before testing for similarity. In this way the psychological similarities between two paranoids will emerge, while the psychological difference between the masochist and his sadistic partner stands out in spite of the great similarity in the *dramatis personae* of their notional worlds viewed uncentred.[20]

The prospect of a rigorous method of notional world comparison—a decision procedure for finding and rating points of coincidence, for instance—is dim. But we have always known that, for the prospects of setting conditions for propositional attitude identity are equally dim. I believe that salt is sodium chloride, but my knowledge of chemistry is abysmal; the chemist believes that salt is sodium chloride as well, but there is not going to be any crisp way of capturing the common core of our beliefs (cf. Dennett 1969). The comparability of beliefs, viewed either as notional attitudes or as propositional attitudes, is not going to be rendered routine by any theoretical stroke. The *gain in precision* one might misguidedly have hoped to obtain by isolating and translating 'the language of thought'—if it exists—would not improve the comparability of *beliefs*, such as mine and the chemist's about salt, but only the comparability of a certain novel kind of sentence—sentences in the head. But sentences are already nicely comparable. The English-speaking chemist and I use exactly the same words to express our belief about salt, and if perchance our brains do as well we will still have the problem of comparability for our beliefs.

A language of thought would give no more leverage in the vexed case of irrational—and especially contradictory—beliefs, and for the same reason. Suppose the hypothesis is bruited that Bill has a particular pair of contradictory beliefs: he believes both that Tom is to be trusted and that Tom is not to be trusted. In any language worth its salt nothing is more cut and dried than determining when one sentence contradicts another, so knowing Bill's language of thought, we search in his brain for the relevant pair of sentences. And we find them! What would this show? The question would still remain: Which (if either) does he believe?

We might find, on further investigation, that one of these sentences was vestigial and non-functional—never erased from the cerebral blackboard, but never consulted either. Or we might find that one sentence (the Mentalese for 'Tom is not to be trusted') was intermittently consulted (and acted upon)—good evidence that Bill believes Tom is not to be trusted, but it keeps slipping his mind. He forgets, and then his natural *bonhomie* takes over and, believing that people in general are to be trusted, he behaves as if he believes Tom is to be trusted. Or perhaps we find truly conflicting behaviour in Bill; he goes on and on in conversation about Tom's trustworthiness, but we note he never turns his back on Tom. One can multiply the cases, filling gaps and extending the extremes, but in none of the cases does the presence or absence of explicit contradiction in the Mentalese play more than a peripheral supporting role in our decision to characterize Bill as vacillating, forgetful, indecisive, or truly irrational. Bill's behaviour counts for more, but behaviour will not *settle* the matter either.

People certainly do get confused and worse; sometimes they go quite mad. To say that someone is irrational is to say (in part) that in some regard he is ill-equipped to deal with the world he inhabits; he ill-fits his niche. In bad cases we may be unable to devise any notional world for him; no possible world would be a place where he fits right in. One could leave the issue there, or one could attempt to be more descriptive of the person's confusion.[21] One could compose an avowedly inconsistent description, citing chapter and verse from the subject's behavioural propensities and internal constitution in support of the various parts of the description. Such an inconsistent description could not be said to be of *a* notional world, since notional worlds, as sets of *possible* worlds, cannot have contradictory properties, but nothing guarantees that a subject has a single coherent notional world. His notional world may be shattered into fragmentary, overlapping, competing worlds.[22] When the theorist, the hetero-phenomenologist or notional world psychologist, takes the course of offering an admittedly inconsistent description of a notional world, it counts not as a settled, positive characterization of a notional world, but as a surrender in the

face of confusion—giving up the attempt at (complete) interpretation. It is analogous to lapsing into direct quotation when
conveying someone's remarks. 'Well, what he *said* was: "Nothing
noths".'

Notional world hetero-phenomenology does not, then, settle
the disputes and indeterminacies, or even sharpen the boundaries
of everyday-folks' thinking about belief; it inherits the problems
and simply recasts them in a slightly new format. One might
well wonder what resources it has to recommend it. The prospect
of constructing the notional world of an actual creature from an
examination of its physical constitution is as remote as can be, so
what value can there be in conceiving of a creature's notional
world? Working in the other direction: starting with a
description of a notional world and then asking how to design a
'creature' with that notional world. Part of the allure of A.I. is
that it provides a way of starting with what are essentially
phenomenological categories and distinctions—features of notional worlds—and working backwards to hypotheses about how
to implement those competences.[23] One starts with representational *powers*, and works towards *means*. Philosophers have also
toyed with this strategy.

The recent philosophical literature on the distinction between
de re and *de dicto* beliefs and other attitudes is replete with sketchy
suggestions for various sorts of mental machinery that might
play a crucial role in drawing that distinction: Kaplan's (1968)
vivid names, the *modes of presentation* of Schiffer (1978) and various
other authors, and Searle's (1979b) *aspects*, to name a few. These
are typically supposed to be definable purely in terms of *narrow
psychology*,[24] so notional attitude psychology should, in principle,
be able to capture them. When we turn to that literature in the
next section, we will explore the prospects for such machinery,
but first there are some more grounds for scepticism about
notional worlds to bring into the open.

The theme of a notional world, a world *constituted* by the mind
or experience of a subject, has been a recurrent leitmotive in
philosophy at least since Descartes. In various forms it has
haunted idealism, phenomenalism, verificationism, and the
coherence theory of truth, and in spite of the drubbing it

typically takes, it keeps getting resurrected in new, improved versions: in Goodman's (1978) *Ways of Worldmaking*, and in Putnam's (1978) recent revaluation of realism, for instance. The ubiquity of the theme is no proof of its soundness in any guise; it may be nothing more than an eternally tempting mistake. In its present guise it runs headlong into an equally compelling intuition about *reference*. If *notional attitudes* are to play the intermediary role assigned to them, if they are to be the counterpart for psychology of Kaplan's concept of the *character* of a linguistic expression, then it should follow that when a psychological subject or creature, with its notional world fixed by its internal constitution, is placed in different contexts, different real environments, this should determine different propositional attitudes for the subject:

notional attitude + environment → propositional attitude

That means that if I and my *Doppelgänger* were to be switched, instantaneously (or in any case without permitting any change of internal state to occur during the transition—the interchange could take a long time so long as I and my *Doppelgänger* were comatose throughout), I would wake up with propositional attitudes *about the things on Twin-Earth*, and my *Doppelgänger* would have propositional attitudes *about things on Earth*.[25] But that is highly counter-intuitive (to many people, I discover, but not to all). For instance, I have many beliefs and other attitudes about my wife, a person on Earth. When my *Doppelgänger* first wakes up on Earth after the switch, and thinks 'I wonder if Susan has made the coffee yet', *surely* he isn't thinking thoughts about *my wife*—he has never met her or heard of her! His thoughts, surely, are about *his* Susan, light years away, though he hasn't an inkling of the distance, of course. The fact that he'd never know the difference, nor would anyone else except the Evil Demon that pulled off the switch, is irrelevant; what no one could *verify* would nevertheless be *true*; his thoughts are not about my wife—at least not until he has had some causal commerce with her.

That is the essence of the causal theory of reference (see, e.g., Kripke 1972; Evans 1973; Donnellan 1966, 1970, 1974) and the thought-experiment nicely isolates it. But intuitions provoked in such wildly science-fictional circumstances are a poor test. Consider the same issue as it could arise in a perfectly possible train of events right here on Earth. In Costa Mesa, California there is, or at any rate used to be, an establishment called Shakey's Pizza Parlor, a garish place featuring an ill-tuned player piano with fluorescent keys, and with various 'funny' hand-painted signs on the walls: 'Shakey's has made a deal with the bank: we don't cash checks, and the bank doesn't make pizza', and so forth. Oddly enough, *very* oddly in fact, in Westwood Village, California, some fifty miles away, there was another Shakey's Pizza Parlor, and it was eerily similar: built to the same blueprint, same ill-tuned player piano, same signs, same parking lot, same menu, same tables and benches. The obvious practical joke occurred to me when first I noticed this, but sad to say I never carried it out. It could easily have been done, however, so let me tell you the tale as if it actually happened.

The Ballad of Shakey's Pizza Parlor

Once upon a time, Tom, Dick and Harry went to Shakey's in Costa Mesa for beer and pizza, and Dick and Harry played a trick on Tom, who was new to the area. After they had ordered their food, and begun eating, Tom went to the men's room whereupon Dick slipped a mickey into Tom's beer. Tom returned to the table, drained his mug and soon fell sound asleep at the table. Dick gathered up the uneaten pizzas, Harry got Tom's hat off the peg behind his head, and then they dragged Tom out to the car and sped to Westwood Village, where they re-established themselves, with a new pitcher of beer and some mugs, at the counterpart table. Then Tom woke up. 'I must have dozed off', he commented, and the evening proceeded, noisily, as before. The conversation turned to the signs and other decorations, and then to graffiti; to the delight of Dick and Harry, Tom pointed towards the men's room and confessed that although he isn't really that sort of guy, tonight he was inspired to carve his initials on the door of the leftmost stall in that men's room. Dick and Harry doubted his word, whereupon Tom offered a wager. He announced he was prepared to bet that his initials were carved on that door.

Dick took the wager, with Harry to referee, and a paper and pencil were produced, on which the explicit expression of the proposition at issue was to be written. At this point, the suspense was high, for whether or not Dick won the wager depended on the exact wording. If Tom wrote: 'I wager $5 that on the leftmost stall door in the men's room in the Costa Mesa Shakey's my initials appear', Tom would win the wager. But if Tom wrote 'I wager $5 that my initials appear on the leftmost stall of the men's room of the pizza parlor in which we currently are seated'—or words to *that* effect—Dick would win. A third possibility was that Tom would compose a sentence that *failed to express a proposition* because it contained a vacuous name or vacuous description: 'the Costa Mesa Shakey's, in which we are now sitting', or 'the men's room of the place wherein I bought and entirely consumed an anchovy pizza on the night of February 11, 1968'. In that case Harry would be forced to declare the wager ill-formed and return the stakes. (If Harry is a strict Russellian about definite descriptions, he may declare Tom's sentence false in these instances, and award the money to Dick.)

But Tom played into their hands, committed himself on paper to the Westwood Village door (though not under *that* description, of course), and lost the bet. The practical joke was explained to him, and Tom, though he admitted he'd been tricked, agreed that he had committed himself to a false proposition and had fairly lost the bet. But which door had he 'had in mind?' Well, in some regards, he could rightly insist, he'd had the Costa Mesa door in mind. He'd vividly recalled the episode of his pen-knife digging into that door. But also he'd vividly 'pictured' the door as being just a few feet away, and eagerly anticipated, in his imagination, the triumph to ensue when the three of them walked into the adjacent men's room to settle the bet. So there was also a lot to be said in favour of his having had the Westwood Village door in mind. Such a puzzle! Clearly this was a job for sober philosophers with a technical vocabulary at their disposal!

There is a distinction, the philosophers say, between belief *de re* and belief *de dicto*. Everyone knows this distinction in his heart, but like many important philosophical distinctions, it is hard to characterize precisely and uncontroversially. We're working on it. In the meantime, we mark the distinction, which tends to get lost in the ambiguities of casual talk, by always using the awkward but at least arguably grammatical 'of' style of attribution, when speaking of beliefs *de re*, reserving the 'that' style for attributions of belief *de dicto*.[26] Thus

(1) Bill believes *that* the captain of the Soviet Ice Hockey team is a man

but it is not the case that

(2) Bill believes *of* the captain of the Soviet Ice Hockey Team that he is a man

since Bill is utterly unacquainted with that stalwart Russian, whoever he is. In contrast,

(3) Bill believes *of* his own father that he is a man.

Surely we all know *that* distinction, the distinction ostended by the example, so now we can proceed to apply it in the case of Shakey's Pizza Parlor. In virtue of Tom's rich causal intercourse with the Costa Mesa Shakey's and the things within it, Tom is entitled to *de re* beliefs relating him to those things. When he wakes up in Westwood Village, as his eyes dart about the room he swiftly picks up the obligatory causal relations with *many* of the objects in Westwood Village as well, including the Westwood Village Shakey's itself. Thus we can catalogue some of the true and false *de re* beliefs of Tom shortly after waking:

Tom believes *of* the Costa Mesa Shakey's:

TRUE	FALSE
that he bought a pizza in it tonight	that he is in it now
that he dozed off in it	that he woke up in it
that he put his hat on a peg in it	that his hat is on a peg in it

Tom believes *of* the Westwood Village Shakey's:

TRUE	FALSE
that he is in it now	that he bought a pizza in it tonight
that he woke up in it	that he dozed off in it
that his hat is on a peg in it	that he put his hat on a peg in it

Where in the normal course of things a person would have a single list of *de re* beliefs, Tom, because of the trick dislocation we have produced, has a dual list of *de re* beliefs; every true *de re* belief has a false twin about a different object. Of course this doubling up of Tom's beliefs is entirely unrecognized by Tom; there is still *something unitary* about his *psychological state*. (We could say: there is unity in his notional world, where there is duality in the real world. Each single notional attitude of his spawns a pair of propositional attitudes, given his peculiar circumstances.)

But problems emerge when we attempt to continue the list of Tom's *de re* beliefs. Tom presumably believes *of* his hat both that it is on a peg in Costa Mesa and that it is on a peg right behind his head. Having noticed the Costa Mesa peg, however casually, Tom can also be said to believe *of* the Costa Mesa peg that his hat is on it (or, putting the two together: he believes of his hat and that peg that the former is on the latter). But can he believe of the peg behind his head, with which his only causal interaction to date has been an infinitesimally weak mutual gravitational attraction, that his hat is on it? The causal theorist must deny it. One would think that Tom's psychological state *vis-à-vis* his hat and its location was quite simple (and so it seems to Tom) but in fact it is quite wonderfully complex, when subjected to philosophical analysis. Tom believes *that* his hat is on a peg behind his head (and that is a true belief): he also believes truly *that* the peg behind his head on which his hat is resting is made of wood. He does not believe *of* that peg, however, that it is made of wood or behind his head. Moreover, Tom believes truly *that* there is a leftmost stall door in the adjacent men's room, and believes falsely *that* the leftmost stall door in the adjacent men's room has his initials carved on it. So he believes that the leftmost stall door has his initials on it, but he does not believe of the leftmost stall door that it has his initials on it.

Some philosophers would disagree. Some (e.g., Kaplan 1968) would say Tom's rapport with the Costa Mesa peg was too *casual* (though causal) to qualify Tom for *de re* beliefs about it. Pushing in the other direction, some (e.g., Kaplan 1978) would hope to weaken the causal requirement (and replace it with something else, still to be determined) so that Tom could have *de re* beliefs about the unseen peg and the unmarked door. And some would stick to their guns, and claim that the admittedly bizarre distinctions drawn in the preceding paragraph were nothing more than the tolerable implications of a good theory put *in extremis* by highly unusual conditions.

The point of pursuing these disagreements, of settling the philosophical dispute, might well be lost on a psychologist. It is tempting to hold that the *philosophical* problems encountered here, even if they are serious, real problems whose solution is worth pursuing, are not problems for *psychology* at all. For note that the different schools of thought about Tom's *de re* beliefs fail to differ in the predictions they would make of Tom's behaviour under various circumstances. Which sentences he can be enticed to bet on, for instance, does not depend on which *de re* beliefs he *really* has. No school can claim predictive superiority based on its more accurate catalogue of Tom's beliefs. Those who hold he does not have any *de re* beliefs about the unseen door will *restrospectively* describe those cases in which Tom makes a losing bet as cases in which he willy-nilly asserts something he does not mean to assert, while those of the opposite persuasion will count him on those occasions as having (willy-nilly) expressed exactly what he believed. In the imagined case, if not perhaps in other, more normal cases, the presence or absence of a particular *de re* belief plays no predictive, hence no explanatory, role. But if in the imagined case it plays no role, should we not abandon the concept in favour of some concept which can characterize the crucial variables in both the normal and the abnormal cases?

The apparent failure of the philosophical distinctions to mesh with any useful psychological distinction may be due, however, to our looking in the wrong place, focussing too narrowly on a contrived *local* indiscernibility and thus missing an important psychological difference that emerges somehow in a broader context. The family of outlandish cases concocted by participants in the literature, involving elaborate practical jokes, tricks with mirrors, people dressed up like gorillas, identical twins and the rest of the theatrical gimmicks designed to produce cases of *mistaken identity*, succeed in producing only momentary or at best unstable effects of the desired sort. One cannot easily *sustain* the sort of illusion required to ground the anomalous verdicts or other puzzles. Drawing verdicts based on short-lived anomalies in a person's psychological state provides a seriously distorted picture of the way people are related to things in the world; our capacity to keep track of things through time is not well described by any theory that atomizes psychological processes into successive moments with certain characteristics.[27] This is all, I think, very plausible, but what conclusion should be drawn from it? Perhaps this: formal semantics requires us to *fix* an object to be evaluated, at a particular time and in a particular context, for truth-value or reference, and while overt linguistic behaviour provides the theorist with candidate objects—utterances—for such a role, *moving the game inside* and positing analogous 'mental' objects or states for such fixing must do violence to the psychological situation. Anyone who imports the categories required for a formal semantic theory and presses them into service in a psychological theory is bound to create a monster. Such a conclusion is, as James might say, 'vagueness incarnate'. In particular, it is not yet clear whether it might be so strong a conclusion as to threaten *all* versions of 'mental representation' theory, all theories that suppose there are syntactic objects in the head for which a principled semantic interpretation can be given.

I find it very difficult to put a crisper expression to this worry, but can for the moment render it more vivid with the aid of an analogy. One of the most inspired skits featured regularly on the television show *Laugh-In* was 'Robot Theater', in which Arte Johnson and Judy Carne played a pair of newly-wed robots. They would appear in some mundane circumstances, making breakfast or hubby-home-from-the-office, and would move about in a slightly jerky simulacrum of human action. But things never worked quite right; Arte would reach out to open a door, grasp just short of the knob, turn wrist, swing arm, and crash headlong into the still closed door; Judy would pour coffee for Arte, but the coffee would miss the cup—no matter, since Arte would not notice, and would 'drain' the cup, turn lovingly to Judy and say 'Delicious!' And so forth. The 'problem', one saw, was that their *notional worlds didn't quite 'match' the real world*; one had the impression that if they had been moved over about a half an inch before they were 'started', everything would have gone swimmingly; *then* their beliefs would have had a chance of being *about* the things in the world around them. Their behaviour was to be explained, presumably, by the fact that each of them contained an internal representation of the world, by consultation with which they governed their behaviour. That's how robots work. This internal representation was constantly updated, of course, *but not continuously*. Their perceptual machinery (and their internal records of their actions) provided them with a succession of snapshots, as it were, of reality, which provoked revisions in their internal representations, but not fast enough, or accurately enough, to sustain a proper match of their notional world, their world-as-represented, and the real world.[28] Hence the behavioural infelicity. The 'joke' is that *we are not at all like that.*

Well, are we or aren't we? The hope of 'cognitive science' is that we *are* like that, only much, much *better*. In support of this conviction, cognitive science can point to precisely the anomalous cases envisaged in the literature on *de re* and *de dicto* beliefs: these are nothing more than experiments, in effect, that *induce pathology* in the machinery, and hence are rich sources of clues about the design principles of that machinery. That one must work so hard to contrive cases of actual pathology just shows how *very* good we are at updating our internal representations. The *process* of keeping track of things is practically continuous, but it will still have a perspicuous description in terms of swift revision of an internal model. Besides, as the pathological cases often show, when one adds *verbal* informing to purely *perceptual* informing of the system, the possibilities of serious dislocation, the creating of notional objects with no real counterparts, and the like, are dramatically increased. Beliefs acquired through the medium of language create problems when they must be meshed with perceptually induced beliefs, but these are problems soluble within the domain of cognitive science.

This suggests that the problems encountered in the story of Shakey's Pizza Parlor come from the attempt to apply a single set of categories to two (or more) very different styles of cognitive operation. In one of these styles, we do have internal representations of things in the world, the content of which in some way guides out behaviour. In the other style we have something like procedures for keeping track of things in the world, which permit us to minimize our *representations* of those things by letting us consult the things themselves, rather than their representatives, when we need more information about them. Reflections on this theme are to be found in the literature of philosophy (e.g., Burge 1977; Kaplan 1968, 1978, forthcoming; Morton 1975; Nelson 1978), psychology (e.g., Gibson 1966; Neisser 1976) and Artificial Intelligence (e.g., Pylyshyn 1979), but no one has yet succeeded in disentangling the goals and assumptions of the various different theoretical enterprises that converge on the topic: the semantics of natural language, the semantics and metaphysics of modal logic, the narrow cognitive psychology of individuals, the broad or naturalistic psychology of individuals in environments and social groups. Armed, tentatively, with the idea of notional worlds, which provides at least a picturesque, if not demonstrably sound, way of describing those matters that belong within the domain of narrow psychology, and distinguishing them from matters requiring a different perspective, perhaps some progress can be made by considering the origins of the problematic distinctions in the context of the theoretical problems that gave birth to them.

V. *De re* and *de dicto* dismantled

Unless we are prepared to say that 'the mind cannot get beyond the circle of its own ideas', we must recognize that some of the things in the world may in fact become objects of our intentional attitudes. One of the facts about Oliver B. Garrett is that he once lived in Massachusetts; another is that the police have been seeking him for many years; another is that I first learned of his existence in my youth; and another is the fact that I believe him still to be in hiding. [Chisholm 1966.]

These are facts *about* Oliver B. Garrett, and they are not trivial. In general, the relations that exist between things in the world in virtue of the beliefs (and other psychological states) of believers are relations we have very good reasons to talk about, so we must have *some* theory or theories capable of asserting that such relations hold. No methodologically solipsistic theory will have that capacity, of course.

The same conclusion is borne in on Quine, the founding father of the contemporary literature on the so called *de re* and *de dicto* distinction, and it requires him to give up, reluctantly, the programme of treating *all* attribution of belief (and other psychological states) as 'referentially opaque'. Non-relationality is the essence of Quine's concept of referential opacity; a context in a sentence is referentially opaque if the symbols occurring within it are not to be interpreted as playing their normal role; are not, for instance, terms denoting what they normally denote, and hence cannot be bound by quantifiers. Frege maintained a similar view, saying that terms in such contexts had an *oblique* occurrence, and referred not to their ordinary denotations, but to their *senses*. Quine's ontological scruples about Fregean senses and their many kin (propositions, concepts, intensions, attributes, intentional objects, ...) force him to seek elsewhere for an interpretation of the semantics of opaque contexts. (See, e.g., Quine 1960, p. 151.) In the end, he handles the myriads of different complete belief-predicates (one for each attributable belief) by analogy with *direct quotation*; to have a belief (non-relationally construed) is to be related to no object or objects in the world save a closed sentence. To believe is to be in an otherwise unanalysed state captured by a lumpy predicate distinguished from others of its kind by containing an inscription of a sentence which it in effect quotes.

> we might try using, instead of the intensional objects, the sentences themselves. Here the identity condition is extreme: notational identity. . . . The plan has its recommendations. Quotation will not

fail us in the way class abstraction did. Moreover, conspicuously opaque as it is, quotation is a vivid form to which to reduce other opaque constructions. [1960, p. 212. See also p. 216, and Quine 1969.]

Such lumpy predicates are not much use, but Quine has long professed his scepticism about the possibility of making any sense of the refractory idioms of intentionality,[29] so he needs opacity only to provide a quarantine barrier protecting the healthy, extensional part of a sentence from the infected part. What one gives up by this tactic, Quine thinks, is nothing one cannot live without: 'A *maxim of shallow analysis* prevails: *expose no more logical structure than seems useful* for the deduction or other inquiry at hand. In the immortal words of Adolf Meier, where it doesn't itch, don't scratch.' (1960, p. 160.) With persistence and ingenuity, Quine staves off most of the apparent demands for relational construals of intentional idioms, but, faced with the sort of case Chisholm describes, he recognizes an itch that must be scratched.

> The need of cross-reference from inside a belief construction to an indefinite singular term outside is not to be doubted. Thus see what urgent information the sentence 'There is someone whom I believe to be a spy' imparts, in contrast to 'I believe that someone is a spy' (in the weak sense of 'I believe there are spies'). [1960, p. 148.]

This then sets the problem for Quine and subsequent authors: 'Belief contexts are referentially opaque; therefore it is prima facie meaningless to quantify into them; how then to provide for those indispensable relational statements of belief, like "There is someone whom Ralph believes to be a spy"?' (Quine 1956.) Quine is led to acknowledge a distinction between two kinds of belief attribution: *relational* attributions and *notional* attributions, in his terms, although others speak of *de re* and *de dicto* attributions, and Quine acknowledges that it comes to the same thing. (It *does* come to the same thing in the literature, but if Quine had meant what he ought to have meant by 'relational' and 'notional' it would not have come to the same thing, as we shall see.) This sets in motion the cottage industry of providing an adequate analysis of these two different kinds of belief-attribution. Unfortunately, three different strains of confusion are fostered by Quine's setting of the problem, though Quine himself is not clearly prey to any of them, nor of course is he entirely responsible for the interpretation of his views that solidified the confusions in the subsequent literature. First, like Chisholm, Quine is struck by only one variety of important relation between believers and the things of the world: cases in which the believer is related to a particular concrete individual (almost exclusively in the examples in the literature, another *person*) in virtue of a belief; focus on these cases has led to a sort of institutional blindness to the importance of other relations. Second, by following Adolf Meier's advice and avoiding explicitly relational construals except when the situation demanded it, Quine helps create the illusion that there are two different *types of belief*, two different sorts of mental phenomena, and not just two different styles or modes of belief-attribution. Quine's acknowledgement that there are times when one is obliged to make an explicitly relational assertion (and then there are times when one can get by with a merely notional assertion) becomes transformed into an imagined demonstration that there are two different sorts of belief: relational beliefs and non-relational belief. Third, putting the first two confusions together, the identification of relational beliefs as beliefs about particular single individuals leaves it tempting to conclude that *general* beliefs (beliefs that are not, intuitively, about any *one* particular thing) are an entirely non-relational variety of beliefs. This subliminal conclusion has permitted an unrecognized vacillation or confusion about the status of general beliefs to undermine otherwise well-motivated projects. I shall treat these three sources of confusion in turn, showing how they conspire to create spurious problems and edifices of theory to solve them.

Chisholm draws our attention to some interesting facts about Garrett, and Quine acknowledges the 'urgent information' imparted by the assertion that there is someone Ralph believes to be a spy.[30] But consider as well a rather different sort of interesting and important fact.

(1) Many people (wrongly) believe that snakes are slimy.

This is a fact about people, but also about snakes. That is to say,

(2) Snakes are believed by many to be slimy.

This is a property that snakes have, and it is about as important a property as their scaliness. For instance, it is an important *ecological* fact about snakes that many people believe them to be slimy; if it were not so, snakes would certainly be more numerous in certain ecological niches than they are, for many people try to get rid of things they think to be slimy. The ecological relevance of this fact about snakes does not 'reduce' to a conjunction of cases of particular snakes being mistakenly believed to be slimy by particular people; many a snake has met an untimely end (thanks to snake traps or poison, say) as a result of someone's *general* belief about snakes, without ever having slithered into rapport with its killer. So the relation snakes bear to anyone who believes *in general* that snakes are slimy is a relation we have reason to want to express in our theories. So too is the relation any particular snake (in virtue of its snakehood) bears to such a believer. Here are some other interesting facts chosen to remind us that not all belief is about particular people.

(3) Snow is believed by virtually everyone to be cold.
(4) It is a fact about charity that some think it superior to faith and hope.
(5) Not having many friends is believed by many to be worse than not having any money.
(6) Democracy is esteemed over tyranny.

What the Quinian quarantine of opaque construal tends to hide from us is that you really cannot make these claims (in a form that allows you to use them in arguments in the obvious ways) unless you can make them in a way that permits explicit relations to be expressed. It will help us to consider a single case in more detail.

Sam is an Iranian living in California, and Herb believes that all Iranians in California should be deported immediately, but he doesn't know Sam from Adam, and in fact has no inkling of his existence, though he knows there are Iranians in California, of course. Supposing Herb is an authority or even just an influential citizen, this belief of his is one that Sam, who enjoys California, would regret. A world in which people have this belief is worse, for Sam, than a world in which no one has this belief. Let us say Sam is *jeopardized* by this belief of Herb's. Who else is jeopardized? All Iranians living in California. Suppose Herb also believes that all marijuana smokers should be publicly whipped. Does this belief also jeopardize Sam? It depends, of course, on whether Sam is a marijuana smoker. Now something follows from

(7) Sam is an Iranian living in California

and

(8) Herb believes all Iranians living in California should be deported immediately.

that does not follow from (7) and

(9) Herb believes all marijuana smokers should be publicly whipped.

What follows is something that licenses the conclusion that Sam is jeopardized by Herb's belief cited in (8), but we can agree that what follows is *not*

(10) Herb believes *of Sam* that he should be deported immediately.

That is, no belief of the sort that impresses Quine and Chisholm follows from (7) and (8). Rather, we are looking for something more like

(11) Sam is a member of the set of Iranians in California, and *of that set* Herb believes that all its members should be deported immediately.

Some might shudder at the idea of having sufficient *acquaintance* with a set to enable one to have *de re* beliefs about it, but sets in any case will not do the job. Sam, learning of his jeopardy, may wish to get out of it, e.g., by leaving California *or* by changing Herb's belief. His leaving California changes the membership of the relevant set—changes which set is relevant—but surely does not alter Herb's belief. In short, it is not set-membership that jeopardizes Sam, but attribute-having.

> (12) There is an attribute (Iranicalifornihood) such that Sam has it, and Herb believes *of* it that anything having it should be deported immediately.

Not just quantifying in, but quantifying over attributes! One might attempt to soften the ontological blow by fiddling around with circumlocutory devices along the lines of

> (13) $(\exists a)(\exists c)$ (a denotes c & Sam is a member of c & Herb believes-true $\ulcorner(x)$ (x is a member of $a \supset x$ should be deported immediately)$\urcorner$)

but they won't work without *ad hoc* provisos of various sorts. If one is going to take belief-talk seriously at all, one might as well let it all hang out and permit quantification over attributes and relations, as well as individuals and classes, in all positions within belief-contexts.[31] See Wallace (1972) for a proposal in a similar spirit.

The brief for this course has been before us from the outset, for it is an implication of the Putnamian arguments that even general propositional attitudes, if they are genuinely *propositional* in meeting Frege's first two conditions, must imply a relation between the believer and things in the world. *My* belief that all whales are mammals is *about whales*; my *Doppelgänger's* counterpart belief might not be (if his Twin-Earth had large *fish* called 'whales', for instance). It is whales I believe to be mammals, and I could not truly be said to believe *that* whales are mammals unless it could also be said of whales that I believe them to be mammals.

Should we say entirely general beliefs are *about anything*? One can read our Fregean condition (b) as requiring it, but there are ways of denying it.[32] For instance, one can note that the logical form of general beliefs such as the belief that all whales are mammals is $(x)(Fx \supset Gx)$, which says, in effect, each thing is such that if it is a whale, it is a mammal. Such a claim is as much about cabbages and kings as about whales. By being about everything, it is about nothing. (Cf. Goodman 1961; Ullian and

Goodman 1977; Donellan 1974.) This does little to still the intuition that when someone believes whales are fishes he is *wrong about whales*, not *wrong about everything*. But here is another troublesome challenge: if general beliefs are always about the things mentioned in their expression, what is the belief that there are no unicorns about? Unicorns? There aren't any. If we are prepared, as I have urged we should be, to quantify over attributes, we can say this belief is about *unicornhood*, to the effect that it is nowhere instantiated.[33] If there were unicorns, the belief would be a false belief about unicorns in just the same way the belief that there are no blue whales is (today) a false belief about blue whales. Pointing to a pod of blue whales we could say: these creatures are believed not to exist by Tom.

In any event a tug-of-war about 'about' does not strike me as a fruitful strategy to pursue (cf. Donnellan 1974), and yet a bit more must be said about this troublesome but practically indispensable term—mentioned once and used twice within this very sentence. 'The term "about" is notoriously vague and notoriously not to be confused with "denotes".' (Burge 1978, p. 128.) Perhaps there is a sense of 'about' that must be carefully distinguished from 'denotes'—much of the literature on *de re* belief is after all a quest to explicate just such a *strong* sense of 'about'—but there is undeniably another, weaker (and in fact much clearer) sense of 'about' of which denotation is the essence. Suppose I believe the shortest spy is a woman. (I have no one in mind, as one says, but it seems a good bet.) Now since I have no one in mind, as one says, there may indeed be a sense in which my belief is not *about* anyone. Yet my belief is either true or false. Which it is depends on the gender of some actual person, the shortest spy, whoever that is. That person is the verifier or falsifier of my belief, the satisfier of the definite description I used to express the belief. There is *that* much relation between me and the shortest spy in virtue of my belief, and paltry and uninteresting as some might find that relation, it is a relation I bear to just one person in virtue of my belief, and we could hardly give it a better name than aboutness—*weak* aboutness, we may call it, allowing for the possible later discovery of stronger, more interesting sorts of aboutness. Suppose Rosa Klebb is the

shortest spy. Then it might be *misleading* to say I was thinking
about her when it occurred to me that the shortest spy was
probably a woman, but in this weak sense of 'about' it would
nevertheless be *true*.[34]

The second confusion, of which Quine himself is apparently
innocent, is promoted by some misdirection in his 'Quantifiers
and Propositional Attitudes' (1956), the chief problem-setter for
the subsequent literature. Quine illustrates the difference between
two senses of 'believes' with a brace of examples.

> Consider the relational and notional senses of believing in spies:
>
> (14) $(\exists x)$ (Ralph believes that x is a spy),
> (15) Ralph believes that $(\exists x)$ (x is a spy).
>
> ... The difference is vast; indeed, if Ralph is like most of us, (15) is
> true and (14) is false. [p. 184. I have changed Quine's numbering of
> his examples to conform to my sequence.]

The examples certainly do illustrate two different attribution
styles: Quine's (14) is ontologically committing where (15) is
reticent; but they also illustrate very different sorts of psychological state; (14) is about what we may call a *specific* belief, while
(15) is about a *general* belief (cf. Searle 1979b). It has proven
irresistible to many to draw the unwarranted conclusion that
(14) and (15) illustrate a relational kind of belief and a notional
kind of belief respectively. What people ignore is the possibility
that *both* beliefs can be attributed in *both* the relational and
notional styles of attribution. On the one hand, some *other*
relational claim (other than (14)) may be true in virtue of (15),
e.g.,

> (16) $(\exists x)$ (x is spyhood & Ralph believes x to be instantiated),

and on the other hand perhaps there are other merely notional
readings (other than (15)) that follow from (14). Many have
thought so; all who have tried to isolate a '*de dicto* component' in
de re belief have thought so, in effect, though their efforts have
typically been confounded by misundertandings about what *de
dicto* beliefs might be. In contrasting (14) with (15), Quine is
comparing apples and oranges, a mismatch concealed by the
spurious plausibility of the idea that the lexically simple change

effected by moving the quantifier takes you back and forth between a relational claim and its closest notional counterpart. The true notional counterpart claim of the relational

(17) $(\exists x)$ (x is believed by Tommy to have filled his stocking with toys)

is not

(18) Tommy believes that $(\exists x)$ (x has filled his stocking with toys),

but something for which we do not have a formal expression, although its intended force can be expressed with scare-quotes:

(19) Tommy believes 'of Santa Claus' that he has filled his stocking with toys.[35]

Tommy's notional world is inhabited by Santa Claus, a fact we want to express in describing Tommy's current belief-state without committing ourselves to the existence of Santa Claus, which the relational claim (17) would do. That is the job for which we introduced notional world talk. Similarly, but moving in the other direction, when we want to distinguish my general belief that all water is H_2O from its notional twin in my *Doppelgänger*, we will have to say, relationally, that *water* is what I believe to be H_2O.

This is not meant in itself to show that there are two importantly different sorts of belief, or more than two, but only that a certain familar alignment of the issues is a misalignment. A further particularly insidious symptom of this misalignment is the myth of the non-relationality of 'pure *de dicto*' beliefs. (What could be more obvious, one might say: if *de re* beliefs are relational beliefs, then *de dicto* beliefs must be non-relational beliefs!) The orthodox position is succinctly summarized—not defended—by Sosa (1970):

(20) 'Belief *de dicto* is belief that a certain *dictum* (or proposition) is true, whereas belief *de re* is belief about a particular *res* (or thing) that it has a certain property.'

One might pause to wonder whether the following definition would be an acceptable substitute:

(21) Belief *de dicto* is belief *of* a certain *dictum* (or proposition) that it is true, whereas belief *de re* is belief *that* a particular *res* (or thing) has a certain property

and if not why not. But instead of pursuing that wonder, I want to focus instead on the wedding of words '*dictum* (or proposition)', and what can be conjured from them. The Latin is nicely ambiguous; it means *what is said*, but does that mean *what is uttered* (the words themselves) or *what is expressed* (what the words are used to assert)? The *OED* tells us that a dictum is 'a saying', which provides yet another version of the equivocation. We saw at the outset that there were those who viewed propositions as sentence-like things and those who viewed them rather as the (abstract) *meanings* of (*inter alia*) sentence-like things. These are importantly different conceptions, as we have seen, but they are often kept conveniently undistinguished in the literature on *de re* and *de dicto*. This permits the covert influence of an incoherent picture: propositions as *mental sayings, entirely internal* goings on that owe their identities to nothing but their intrinsic properties, and hence are entirely non-relational.[36] At the same time, these mental sayings are not mere *sentences*, mere syntactic objects, but *propositions*. The idea is fostered that a *de dicto* propositional attitude attribution is an entirely internal or methodologically solipsistic determination of content that is independent of how the believer is situated in the world. The subconscious metaphysics of this reliance on '*de dicto*' formulations as content-specifiers is the idea that somehow the mental embrace of a *sentence* causes the *proposition* the sentence *expresses* to be *believed*.[37] (Details of the embracing machinery are left to the psychologist. What is mildly astonishing to me is the apparent willingness of many psychologists and theorists in A.I. to accept this setting of their task without apparent misgivings.)

No one is clearer than Quine about the difference between a sentence and a proposition, and yet the legerdemain with which he attempts to spirit propositions off the scene and get by with just sentences may have contributed to the confusion. In *Word and Object* (1960), after introducing the various problems of belief-contexts, and toying with a full-blown system of propositions, attributes, and relations-in-intension to handle them, Quine shows how to renounce these 'creatures of darkness' in favour of the lumpy predicates of direct quotation of sentences. Earlier (1956) he had given a curious defence of this move:

> How, where, and on what grounds to draw a boundary between those who believe or wish or strive that *p*, and those who do not quite believe or wish or strive that *p*, is undeniably a vague and obscure affair. However, if anyone does approve of speaking of belief of a proposition at all and of speaking of a proposition in turn as meant by a sentence, then certainly he cannot object to our semantical reformulation '*w* believes-true *S*' on any special grounds of obscurity; for '*w* believes-true *S*' is explicitly definable in *his* terms as '*w* believes the proposition meant by *S*'. [pp. 192–3.]

This tells us how the believer in propositions, attributes and the like is to make sense of Quine's new predicates, but it does not tell us how Quine makes sense of them. He never tells us, but given the 'flight from intension' he advocates for us all, one can suppose that he doesn't think they really make much sense. They are just a device for getting over a bad bit of material that other people keep insisting on inserting into the corpus of serious utterance. The function of Quine's paraphrases is to permit the imaginary civil servant who translates the corpus into the 'canonical notation' to get to the end of the sentence 'safely', and on to the next sentence—confident that he has made *at least* enough new predicates to ensure that he never uses the same predicate to translate two differently intended propositional attitude claims. If sense is ever made of such claims, they can be recovered without loss of distinctness from canonical deep-freeze. In the meantime, Quine's predicates are notional to a fault, but also inert. The problem with this emerges in three quotations:

(A) In the ... opaque sense 'wants' is not a relative term relating people to anything at all, concrete or abstract, real or ideal. [1960, pp. 155–6.]

(B) If belief is taken opaquely then [*Tom believes that Cicero denounced Catiline*] expressly relates Tom to no man. [p. 145.]

(C) (1) 'Tully was a Roman' is trochaic.

(2) The commissioner is looking for the chairman of the hospital board.

Example (2), even if taken in the not purely referential way, differs from (1) in that it still seems to have far more bearing on the chairman of the hospital board, dean though he be, than (1) has on Tully. Hence my cautious phrase 'not purely referential', designed to apply to all such cases and to affirm no distinction among them. If I omit the adverb, the motive will be brevity. [p. 142.]

(A) is unbending: there is *nothing* relational in the opaque sense of 'want'; in (B) the opaque claim does not *expressly* relate Tom to anyone; in (C) it is acknowledged that example (2) 'has a bearing on' a particular person, and hence is to be taken as somewhat referential, but not purely referential. In fact, in all these cases there is clear evidence of what we might call *impure referentiality*. Wanting a sloop 'in the notional sense' does not relate one pre-eminently to any particular sloop, but this wanting relief from slooplessness, as Quine calls it, nevertheless has a bearing on sloops; if no one wanted relief from slooplessness, sloops would not be as numerous, or expensive, as they are. One cannot believe *that* Cicero denounced Catiline without believing *of* Cicero that he denounced Catiline. And it is a fact about the chairman of the hospital board that he is being sought by the commissioner, in virtue of the truth of (2).

This impure referentiality has been noted by several authors, and Quine himself provides the paradigm for the later analyses with his example

(22) Giorgione was so-called because of his size (1960, p. 153)

in which, as Quine notes, 'Giorgione' does double-duty; the perspicuous expansion of (22) is

(23) Giorgione was called 'Giorgione' because of his size.

Castañeda (1967), Kiteley (1968), Loar (1972), and Hornsby (1977) develop the theme of the *normally* dual role of singular terms within the clauses of sentences of propositional attitude— for instance to explain the apparent role of the pronoun in such sentences as

(24) Michael thinks that that masked man is a diplomat, but he obviously is not. (Loar 1972, p. 49.)

This gets part of the story right, but misses the extension of the doctrine from singular terms to general terms. When we quantify over attributes, as in (12) and (16), we complete the picture. A general term within a propositional clause in a sentence of propositional attitude *normally* plays a dual role—as suggested by such sentences as

(25) Herb believes that all Iranians in California should be deported, but none of them have anything to fear from him.[38]

Once we distinguish notional from relational attributions, and distinguish *that* distinction from the distinction between specific and general beliefs, we can see that many of the claims that have been advanced in the effort to characterize '*de re* belief' apply across the board and fail to distinguish a special sort of belief. First, consider the distinction between notional and relational attributions, which emerges when we tackle the problem of relating a subject's narrow-psychological states to his broad, propositional states. We start with the myth that we have determined the subject's notional world, and now must line up his notional attitudes with a set of propositional attitudes. This is a matter of looking for the best fit. As we have seen, in the case of seriously confused subjects, there may be no good fit at all. In the case of an irrational subject, there will be no flawless fit because there is no *possible* world in which the subject could be happily situated. In the case of a badly misinformed subject, there will simply be an area of mismatch with the *actual* world. For instance, the child who 'believes in Santa Claus' has Santa Claus in his notional world, and no one in the real world is a suitable counterpart, so some of the child's notional attitudes about Santa Claus (Santa-Claus-about) cannot be traded in for *any* propositional attitudes at all. They are like sentence-tokens whose character is such that in the context in which they occur they fail to determine *any* content. As Donellan claims, when such a child *says* 'Santa Claus is coming tonight', what he says does not express any proposition (1974, p. 234). We can add: the child's psychological state at that time is a notional attitude that determines no propositional attitude. McDowell (1977) and Field (1978) also endorse different versions of this claim, which can be made to appear outrageous to anyone who insists on viewing beliefs as (nothing but) propositional attitudes. For one must say in that case that no one has beliefs about Santa Claus, and no one could; some people just think they do! Santa-believers have many propositional attitudes, of course, but also some psychological states that are not any propositional attitudes at all. What do they have in their place? (Cf. Blackburn 1979.) Notional attitudes. The attribution of notional attitudes to the child who believes in Santa Claus will provide us with all the understanding and all the theoretical leverage we need to account for the child's behaviour. For instance, we may derive genuine propositional attitudes from the child's merely notional attitudes when the world occasionally obliges.[39] Tommy believes the man in the department store with the phony beard will bring him presents because he believes that the man is Santa Claus. That's why Tommy tells *him* what he wants.

On other occasions, e.g., at Shakey's Pizza Parlor, there is an embarrassment of riches: too many objects in the real world as candidate *relata* when we go relational and trade in notional attitude claims for propositional attitude claims. This possibility is not restricted to specific beliefs about individuals. Putnam's (1975a) discussions of natural kind terms reveals that the same problem arises when we must decide which general proposition someone believes when he believes something he would express with the words 'All elms are deciduous', when we know he cannot tell an elm from a beech. In such a case different properties or attributes are the candidate *relata* for the beliefs.

The distinction between general and specific beliefs has not yet been made clear, but only ostended, but in the meantime we can note that when we go relational, we will trade in specific notional attitude claims for specific relational—that is, propositional attitude—claims, and general notional attitude claims for general relational—that is, propositional attitude—claims. Ralph's banal belief cited notionally in (15) can also be cited relationally as (16), while Ralph's portentous belief, cited relationally in (14), could also be cited notionally if there were a reason—for instance if Ralph had been duped into his urgent state by pranksters who had convinced him of the existence of a man who never was. The necessary conditions for sustaining a relational attribution are entirely independent of the distinction between specific and general attitudes. This can be seen if we insert examples of general beliefs into the familiar arenas of argument.

(26) Tom believes that whales are mammals.

This is a general belief of Tom's, if any is, but earlier we suggested that Tom cannot truly be said to believe *that* whales are mammals unless we are prepared to say as well *of whales* that they are believed by Tom to be mammals. Formally, such a belief-claim would license *quantifying in* as follows:

(27) $(\exists x)$ (x = whalehood & Tom believes of x that whatever instantiates it is a mammal)

If so, then must Tom not have a *vivid name* of whalehood? (Kaplan 1968.)

Contrast (26) with

(28) Bill believes that the largest mammals are mammals.

If all Bill knows about whales is that they are the largest mammals, he has a very unvivid name of whalehood. Most of us are in closer rapport with whales, thanks to the media, but consider

(29) Tom believes that dugongs are mammals.

There is certainly an intuitive pull against saying that dugongs are believed by Tom to be mammals, for Tom, like most of us, is cognitively remote from dugongs. He could get closer by learning (in the dictionary) that dugongs are of the order Sirenia, large aquatic herbivores that can be distinguished from manatees by their bilobate tail. Now is Tom equipped to believe *of* dugongs that they are mammals? Are you? It is one thing to believe, as a monolingual German might, that the sentence 'Dugongs are mammals' is true; that does not count at all as believing dugongs are mammals; it is another thing to know English and believe that the sentence 'Dugongs are mammals' is true, and hence believe (in a very minimal sense) that dugongs are mammals— this is roughly the state most of us are in.[40] If you have read a book about dugongs, or seen a film, or—even better—seen a dugong in a zoo, had a dugong as a pet, then you are much more secure in your status as a believer that dugongs are mammals. This is not a question of the conditions for believing some particular dugong to be a mammal, but just of believing in general that (all) dugongs are mammals.[41]

Kaplan initially proposed vivid names as a condition of genuine *de re* belief, but maintained that there was no threshold of vividness above which a name must rise for one to believe *de re* with it. He has since abandoned vividness as a condition of *de re* belief, but why was the idea so appealing in the first place? Because vividness, or what amounts to vividness as Kaplan characterized it (1968), is a condition on *all* belief; one must be richly informed about, intimately connected with, the world at large, its occupants and properties, in order to be said with any propriety to have beliefs. All one's 'names' must be somewhat vivid. The idea that some beliefs, *de dicto* beliefs, have no vividness requirement is a symptom of the subliminal view of *de dicto* beliefs as mere mental sayings—not so different from the monolingual German's state when he has somehow acquired 'Dugongs are mammals' as a true sentence.

Yet another theme in the literature concerns the admissibility or inadmissibility of substitution of co-designative expressions within the propositional clauses of belief-attributions. It is said that when one is attributing a belief *de re*, substitution is permissible, otherwise not. But if all general beliefs have relational renderings (in which they are viewed as beliefs *of* certain attributes, in the manner of (12) and (27)), do these admit of substitution? Yes, under the same sorts of pragmatic constraints for securing reference without misleading that have been noted in the literature for beliefs about individuals (Sosa 1970; Lycan and Boër 1975; Hornsby 1977; Searle 1979b—to name a few).

(30) My wife wants me to buy her a sweater the colour of your shirt.

(31) Your grandfather thought that children who behave the way you're behaving should be sent to bed without supper.

and of course Russell's famous

(32) I thought your yacht was longer than it is.

In (30) and (31) the attributes are referred to by description (cf. Aquila 1977, p. 91). Sometimes reference to *objects* is secured by something like direct ostension, often with the aid of demonstratives, and Donnellan (1966, 1968, 1970, 1974), Kaplan (1978, forthcoming) and others have claimed that there is a fundamental difference between *direct* reference and the somehow indirect reference obtainable via definite description. Again postponing consideration of the merits of these claims, note that the presumed distinction fails to mark a special feature of specific belief about particular objects. We can say, in an act of direct reference

(33) *This* is the Eiffel Tower and *that* is the Seine.

We can also say, in the same spirit:

(34) An English horn sounds *thus*, while an oboe sounds *thus*.
(35) To get a good vibrato, do *this*.

A Frenchman says that something has a certain *je ne sais quoi*; what he means is that *il sait* perfectly well *quoi*, he just can't *say quoi*, for the property in question is a *quale* like the taste of pineapple or the way you look tonight. (Cf. Dennett unpublished). He can predicate the property of something only by making an identifying reference to the property by such a definite description, and if this occurs within an intensional context, it often must be read transparently, as in (30) and (31) or

(36) Rubens believed that women who look like you are very beautiful.

With relationality, vividness, substitutability, and direct

demonstrative reference set behind us, we can return to the postponed question: what is the distinction between general and specific belief? We should be able to draw the distinction independently of genuinely relational attributions, hence at the level of notional attitude attributions—since we want to distinguish specific Santa-Claus-about states of mind from general states of mind, for instance. There are two opposing views in the philosophical literature that must be disentangled from their usual involvement with referential and attributive uses of definite descriptions in public speech acts, and with problems arising in attempts to capture a specifically *causal* theory of reference. One can be called the Definite Description view and the other, its denial, which takes various forms, can be noncommittally called the Special Reference view. On the former view, the only distinction to mark is that between believing *all Fs* are *Gs* or *some Fs* are *Gs* on the one hand (general beliefs), and believing *just one F* is *G* on the other (specific belief) where such specific beliefs are viewed as adequately captured by Russell's theory of definite descriptions. The latter view, while not denying the existence of that distinction in logical form, insists that even the beliefs the former view calls specific, the beliefs expressed with Russellian definite descriptions, are properly general, while there is a further category of truly specific beliefs, which are more strongly *about* their objects, because they pick out their objects by some sort of *direct* reference, unmediated by (the *sense* of) descriptions of any sort. To the great frustration of anyone attempting to address the issue as, say, a psychologist or A.I. theorist, *neither* view as typically expressed in the philosophical literature makes much plausible contact with what must be, in the end, the empirical issue: what it is, *literally* in the heads of believers, that makes a psychological state a belief about a particular object. The Definite Description view, if taken at literal face value, would be preposterous sententialism; recognizing that, its adherents gesture in the direction of recognitional mechanisms, procedures, and criterial tests as ways of capturing the effect of definite descriptions in more psychologically realistic (if only because vaguer and hence *less unrealistic*) terms. On the other side, the critics of all such proposals have their own gestures to make in the direction of unspecified causal intimacies and genealogical routes between psychological states and their true objects. Since no progress can be made at those levels of metaphoricalness, perhaps it is best to revert to the traditional philosophical terms (ignoring psychology for a while) to see what the issue might be.

In favour of the Definite Description view it can be noted that a Russellian definite description does indeed manage to cull one specific object from the domain of discourse when all goes well. Against this, the Special Reference view can hold that the Russellian expansion of a definite description reveals its disguised generality: to say that *the man who killed Smith is insane* is, on Russellian analysis, to say that *some* person x has a certain property (some F is G): some x has the property of being identical with any person who killed Smith and being insane. This generality is also revealed to intuition (it is claimed) by the 'whoever' test: 'the man who killed Smith', if it means '*whoever* killed Smith' (as it must, on Russellian analysis) makes no specific reference. Intuitively, 'whoever killed Smith' does not *point at* anyone in particular, but rather *casts a net* into the domain. A genuine specific belief points directly to its object, which is identified not by its being the unique bearer of some property, but by its being . . . the object of the belief. Why should we suppose there are any such beliefs? There are two different motivations entangled here, one metaphysical, having to do with essentialism, and the other psychological, having to do with differences in psychological state that we dimly appreciate but have yet to describe. First we must expose the metaphysical issue, and then, at long last, we can address the psychological issue.

Consider what we should say about the identity conditions of the belief cited in

(37) Tom believes that the shortest spy is a woman.

If we treat this attribution relationally, as we must if we want to know whether Tom's belief is true or false, then the proposition he believes, given the world he is embedded in, is about (in the weak sense) some actual person, Rosa Klebb, let us suppose, so his belief is true. Were someone else, Tiny Tim Traitor, the shortest spy, then in *that* world, Tom would have had a false belief. Would we say, though, that in *that* world Tom would have had the same belief, or a different belief? Here we must be careful to distinguish narrow-psychological state or notional attitude from propositional attitude. We must also be careful to distinguish reflections about what *could have been* the case from what *may become* the case. For note that Tom's narrow-psychological state could stay constant in the relevant regards for, say a year, during which time the 'title' of shortest spy changed hands—entirely unbeknownst to Tom of course. For months Rosa Klebb was shortest spy, until Tiny Tim took the KGB's ruble. For those first few months Tom believed something true, and then, when Tiny Tim took over, Tom believed something false. Different propositions, it must be. In fact, as many different propositions in that year as you like, depending on how finely you slice the specious present. The changes in Tom, however, as each evanescent proposition flashes by, as a series of true propositions (the shortest spy on January 1 is a woman, etc.) is followed, suddenly, by a series of false propositions, are what Geach would call Cambridge changes— like that change that befalls you when you suddenly cease to have the property of being nearer the North Pole than the oldest living plumber born in Utah. Not all notional attitudes are thus related to propositional attitudes, of course.

(38) Tom believes that the youngest member of the Harvard class of 1950 graduated with honours.

Which proposition he believes in this case does not change from day to day; moreover, whoever this belief is about (in the weak sense), it is about that person for all eternity. We cannot say that someone else might in the future *come to be* the object (in the weak sense) of that belief, but *perhaps* we can make sense of the suggestion that, counterfactually, someone else *could have been* the object of *that very belief*. To see just what this involves, let us review the possibilities. Fix Tom's *notional* attitude. First there are possible worlds in which Tom, with his notional attitude or narrow-psychological state, *believes no proposition at all*; these are worlds, for instance, without Harvard, and without any Twin-

Harvard either. Second, there are possible worlds in which Tom believes a proposition not about Harvard but about some Twin-Harvard and its youngest graduate of Twin-1950. Third, there are worlds in which Tom believes the very same proposition he believes in the actual world, but in which someone else is the youngest graduate. The identity of the proposition is tied to the attribute, youngest-graduatehood, not to its bearer. And in (37) the propositions believed are tied to shortest-spyhood, not the bearers of the title. That, at any rate, is the doctrine that must be sustained if (37) and (38) are to be distinguished by the Special Reference view from genuinely specific beliefs. This doctrine requires us to make sense of the claims that it is possible that someone else could have been the shortest spy (on January first, on January second, . . .), and it is possible that someone else could have been the youngest graduate of the Harvard class of 1950. Now consider

(39) Tom believes the person who dropped the gum wrapper (whoever it might be) is a careless slob.

Suppose (since some say it makes a difference) that Tom did not see this person, but just the discarded gum wrapper. Now Tom's belief in this instance is, as it happens, about (in the weak sense) some one individual, whoever dropped the gum wrapper, and the belief will be about that one individual through eternity, but had history been just a bit different, had someone else dropped the gum wrapper, then the very same belief of Tom's (the very same notional attitude *and* the very same propositional attitude) could have been about someone else. Could it have been about someone else if Tom had seen the wrapper being dropped? Why not? What difference could it make if Tom sees the person? If Tom sees the person, no doubt

(40) Tom believes that the person he saw drop the gum wrapper (whoever it might be) is a careless slob

will also be true. This belief is weakly about the same careless individual, but had history been just a bit different, it too would have been about someone else. It is at just this point that the believer in the Special Reference view of *de re* belief intervenes to insist that on the contrary, had history been just a bit different, had Tom seen someone else drop the gum wrapper, *he would have had a different belief.* It would have been a different belief because it would have been about (in the *strong* sense) a different person. It is easy to confuse two different claims here. It is tempting to suppose that what the *de re* theorist has in mind is this: had history been a little different, had a tall, thin person been seen dropping the gum wrapper instead of a short, fat person, Tom's perceptions would have been quite different, so he would have been in quite a different narrow-psychological state. But although this would normally be true, this is not what the *de re* theorist has in mind; the *de re* theorist is not making a claim about Tom's narrow state, but about Tom's propositional attitude: had history been a little different, had the short, fat person's twin brother dropped the gum wrapper under indistinguishable circumstances, Tom's notional attitude could have been just the same, we can suppose, but he would have had a different belief, a different propositional attitude altogether, because—just because—his notional state was directly caused (in some special way) by a different individual.[42]

The metaphysics of the view can be brought into sharp focus with a contrived example. Suppose Tom has been carrying a lucky penny around for years. Tom has no name for this penny, but we can call it Amy. Tom took Amy to Spain with him, keeps Amy on his bedside table when he sleeps, and so forth. One night, while Tom sleeps, an evil person removes Amy from the bedside table and replaces Amy with an imposter, Beth. The next day Tom fondles Beth lovingly, puts Beth in his pocket and goes to work.

(41) Tom believes he once took the penny in his pocket to Spain.

This true sentence about Tom asserts that he has a particular false belief about Beth but had history been a little different, had the evil person not intervened, this very belief would have been a true belief about Amy. But according to the Special Reference theory of *de re* beliefs, there are some other beliefs to consider:

(42) Tom believes of Amy that she is in his pocket.

(43) Tom believes of Amy that he once took her to Spain.

These are beliefs strongly about Amy, the first false, the second true, and their being about Amy is *essential* to them. About them we cannot say: had history been a bit otherwise *they* would have been about Beth. If so, then one must also consider:

(44) Tom believes of Beth that she is in his pocket.
(45) Tom believes of Beth that he once took her to Spain.

These are beliefs strongly about Beth, the first true, the second false, and their being about Beth is *essential* to them. About them we cannot say: had history been a bit otherwise *they* would have been about Amy. Why should we adopt this vision? It cannot be because we need to grant that Tom has beliefs about both pennies, for in (41) his belief is (weakly) about Beth and

(46) Tom believes that the penny he took to Spain is in his pocket

attributes to Tom a belief (weakly) about Amy. Let us grant that in all cases of beliefs that are *weakly* about objects in virtue of those objects being the lone satisfiers of a description, we can insert, to mark this, the phrase 'whoever it might be'. This will often have a *very* odd ring, as in

(47) Tom believes that his wife (whoever she might be) is an excellent swimmer

but only because the pragmatic implication of the inserted phrase suggests a very outlandish possibility—unless it merely is taken to suggest that the *speaker* of (47) is unable to pick Tom's wife out of a lineup.

Is there then another reason we should adopt this view? It cannot be because we need to distinguish a special class of *de re* beliefs strongly about objects in order to mark a distinguished class of cases where we may *quantify in*, for we may quantify in whenever a belief is only weakly about an individual, although such a practice would often be very misleading. It would be very misleading, for instance, to say, pointing to Rosa Klebb during her reign as shortest spy: 'Tom believes her to be a woman'. It would also be very misleading, though, for the evil person to hold up Amy and say 'Tom believes he once took this penny to Spain'. These claims would be misleading because the normal pragmatic implication of a relational claim of this sort is that the believer can identify or reidentify the object (or property) in question. But nothing about the manner of acquisition of a belief could *guarantee* this against all future misadventures, so *any* theory of '*de re* belief' would license potentially misleading relational claims. (Cf. Schiffer 1978, pp. 179, 188.)

If it still seems that there is an elusive *variety of belief* that has not been given its due, this is probably because of a tempting misdiagnosis of the sorts of examples found in the literature. Dragging out a few more examples may serve to lay the ghost of *de re* belief (in this imagined sense that distinguishes it as a subvariety) once and for all. Suppose I am sitting in a committee meeting, and it occurs to me that the youngest person in the room (whoever that is—half a dozen people present are plausible candidates) was born after the death of Franklin D. Roosevelt. Call that thought of mine *Thought A*. Now in the weak sense of 'about', Thought A is about one of the people present, but I know not which. I look at each of them in turn and wonder, e.g., 'Bill, over there—is it likely that Thought A is *about him*?' Call *this* thought of mine *Thought B*. Now *surely* (one feels) Thought B is *about Bill* in a much more direct, intimate, strong sense than Thought A is, even if Thought A does turn out to be about Bill. For one thing, I *know* that Thought B is about Bill. This is, I think, an illusion. There is only a difference in degree between Thought A and Thought B and their relation to Bill. Thought B is (weakly) about whoever is the only person I am looking at and whose name I believe to be Bill and . . . for as long as you like. Bill, no doubt, is the lone satisfier of that description, but had his twin brother taken his place unbeknownst to me, Thought B would not have been about Bill, but about his brother. Or more likely, in that eventuality, I would be in a state of mind like poor Tom's in Shakey's Pizza Parlor, so that *no* psychologically perspicuous rendition of my *propositional* attitudes is available.

Another example, with a different flavour. George is Smith's murderer (as Gracie well knows) and she rushes in to tell him that Hoover thinks he did it. Alarmed, George wants to know how Hoover got on to him. Gracie reasons: Hoover knows that the one and only person who shot Smith with a ·38, left three unrecordable fingerprints on the window, and is currently in possession of the money from Smith's wallet is Smith's murderer; since George is the lone satisfier of Hoover's description, Hoover believes of George that he did it. 'No, Gracie,' George says, 'Hoover knows only that *whoever* fits this description is Smith's murderer. He doesn't know that *I* fit the description so he doesn't know that I am Smith's murderer. Wake me up when you learn that *I* am suspected.' (Cf. Sosa 1970.) George has misdiagnosed his situation, however, for consider the case in which Poirot gathers all the houseparty guests into the drawing room and says 'I do not yet know *who the murderer is*; I do not even have a suspect, but I have deduced that the murderer, whoever he is, is the one and only person in the drawing room with a copy of the pantry key on his person.' Search, identification and arrest follow. It is not true that George is safe so long as Hoover's beliefs are of the form *whoever fits description D is Smith's murderer*, for if description D is something like 'the only person in Clancy's bar with yellow mud on his shoes' the jig may soon be up.[43] One is a (minimal) suspect if one satisfies *any* definite description Hoover takes to pick out Smith's murderer. It follows trivially that Smith's murderer is a minimal suspect (because he satisfies the description 'Smith's murderer') even in the situation when Hoover is utterly baffled, but merely believes the crime has been committed by a single culprit. This would be an objectionable consequence only if there were some principled way of distinguishing minimal suspects from genuine or true or *de re* suspects, but there is not. Thus as Quine suggests, the apparently sharp psychological distinction between

(48) Hoover believes that someone (some *one*) murdered Smith

and

(49) Someone is believed by Hoover to have murdered Smith

collapses (the logical difference in the ontological commitment of the speaker remains). It remains true that in the case in which

Hoover is baffled he would naturally deny to the press that there was anyone he believed to be the murderer. What he would actually be denying is that he knows more than anyone knows who knows only that the crime has been committed. He is certainly not denying that he has a *de re* belief directly about some individual to the effect that he is the murderer, a belief he has acquired by some intimate cognitive rapport with that individual, for suppose Hoover wrestled with the murderer at the scene of the crime, in broad daylight, but has no idea who the person was with whom he wrestled; surely on anyone's causal theory of *de re* belief, that person is believed by him to be the murderer, but it would be most disingenuous for Hoover to claim to have a suspect. (Cf. Sosa 1970, pp. 894ff.)

The tying of belief-identity and belief-reference to a causal condition is a plausible proposal because, in most instances, varying degrees of causal intimacy in the past can be used to distinguish weaker from stronger relations between believers and their objects, such that the stronger relations have implications for future conduct, but in unusual situations the normal implications do not hold. From the fact that we can produce dislocations, we see that the causal requirement is not in itself necessary or sufficient; effects are as important as causes. What is necessary is the creation of a notional object in the subject's notional world. This will not typically happen in the absence of causal commerce; it is unlikely for something to 'get into position' for someone to have beliefs about it without having been in causal interaction of *some* sort with the believer, but we can force this result (the Westwood Village peg, the door, Poirot's culprit) in special cases. The special cases draw attention to an independence in principle—if seldom in practice—between psychologically salient states and their metaphysical credentials.

The believer in *de re* belief must decide whether or not the concept at issue is supposed to play a marked role in behavioural explanations (see, e.g., Morton 1975; Burge 1977). One view of *de re* belief would not suppose that anything at all about Tom's likely behaviour follows from the truth of

> (50) Tom believes *of* the man shaking hands with him that he is a heavily armed fugitive mass murderer.

This view acquiesces in what might be called the psychological opacity of semantic transparency (we may not know at all what our beliefs are about), and while I can see no obstacle to defining such a variety of propositional attitude, I can see no use for such a concept, since nothing of interest would seem to follow from a true attribution of such a belief. Suppose, on this view, that something, *a*, is believed by me to be *F*. It does not follow that *a* is not also believed by me to be *not-F*; and if *a* is also believed by me to be *G*, it does not follow that *a* is believed by me to be *F and G*; it does not even follow from the fact that *a* is believed by me to be *the only F*, that no other object *b* is also believed by me to be *the only F*. The premise of the quest for *de re* belief was that there were interesting and important relations between believers and the objects of their beliefs—relations we had reason to capture in our theories—but this termination of the quest lands us with relations that are of only intermittent and unprojectible interest. That being so, there is no longer any motivation I can see for denying that one believes *of* the shortest spy that he is a spy. The formal apparatus for making such relational claims is available, and once one has asserted that two objects are related, however minimally, as believer and object, one can go on to assert whatever other facts about the believer's state of mind or the object's situation might be relevant to plotting and explaining the relevant future careers of both.

If, shunning this view, one seeks a view of *de re* belief as somehow a psychologically distinguished phenomenon, then it cannot be a theory well named, for it will have to be a theory of distinctions within *notional* attitude psychology. If *a* is an object *in my notional world* that I believe to be *F* it *does* follow that I do not also believe *a* to be *not-F*, and the other implications cited above fall into place as well, but only because notional objects are the 'creatures' of one's beliefs (cf. Schiffer 1978, p. 180). Having created such creatures, we can then see what real things (if any) they line up with, but not from any position of privileged access in our own cases. There is a very powerful intuition that we can have it both ways: that we can define a sort of *aboutness* that is *both* a real relation between a believer and something in the world and something to which the believer's access is perfect. Evans calls this Russell's Principle: *it is not possible to make a judgement about an object without knowing what object you are making a judgement about.* In Russell's case, the attempt to preserve this intutition in the face of the sorts of difficulties encountered here led to his doctrine of knowledge by acquaintance and hence inevitably to his view that we could only judge *about* certain abstract objects or certain of our own internal states. The Principle becomes:

> Whenever a relation of supposing or judging occurs, the terms to which the supposing or judging mind is related by the relation of supposing or judging must be terms with which the mind in question is acquainted. [Russell 1959, p. 221.]

Here the term 'term' nicely bridges the gap, and prepares the way for precisely the sort of theory Chisholm abhors in the opening quotation of this section: a theory that supposes 'the mind cannot get beyond the circle of its own ideas'. The way out is to give up Russell's Principle (cf. Burge 1979a), and with it the idea of a special sort of *de re* belief (and other attitudes) intimately and strongly *about* their objects. There still remains a grain of truth, however, in Russell's idea of a special relation between a believer and some of the *things he thinks with*, but to discuss this issue, one must turn to notional attitude psychology and more particularly to the 'engineering' question of how to design a cognitive creature with the sort of notional world we typically have. In that domain, now protected from some of the misleading doctrines about the mythic phenomena of *de re* and *de dicto* beliefs, phenomenological distinctions emerge that cut across the attempted boundaries of the recent philosophical literature, and hold some promise of yielding insights about cognitive organization.

Since this paper has already grown to several times its initially intended size, I will save a detailed examination of these distinctions for other occasions, and simply list what I consider the most promising issues.

(a) *Different ways of thinking of something.* This is a purely notional set of distinctions—none of the ways entails the existence of something of which one is thinking. Vendler (1976, forthcoming) has some valuable reflections on this topic (along with his unintended *reductio ad absurdum* of the causal theory of reference for belief).

(b) *The difference between (episodic) thinking and believing.* At any time we have *beliefs* about many things we are unable to *think* about not because the beliefs are unconscious or tacit or unthinkable, but just because we are temporarily unable to *gain access* to that to which we may want to gain access. (Can you *think* about the person who taught you long division? If you can 'find the person in your memory' you will no doubt discover a cache of beliefs about that person.) Any plausible psychological theory of action must have an account of how we recognize things, and how we keep track of things (cf. Morton 1975), and this will require a theory of the strategies and processes we use in order to exploit our beliefs in our thoughts.

(c) *The difference between explicit and virtual representation.* (Dennett forthcoming a.) When I hop into a rental car and drive off, I expect the car to be in sound working order, and hence I expect the right front tyre to have a safe tread—I would be *surprised* if I discovered otherwise. Not only have I not *consciously* thought about the right front tyre; I almost certainly have not *unconsciously* (but still explicitly) represented the tread of the right front tyre as a separate item of belief. The difference that *having attended to something* (real or merely notional) makes is not the difference between *de re* and *de dicto*; it is a real difference of some importance in psychology.

(d) *The difference between linguistically infected beliefs and the rest*—what I call opinions and beliefs in 'How to Change Your Mind' (1978a). No dog could have the opinion that it was Friday, or that the shortest dog was a dog. Some people have supposed that this means that animals without language are incapable of *de dicto* belief. It is important to recognize the independence of this distinction from the issues at stake in the *de re/de dicto* debates.

(e) *The difference between artefactual or transparently notional objects and other notional objects.* We can conjure up an imaginary thing—just to daydream about, or in order to solve a problem, e.g., to design a dream house or figure out what kind of car to buy. It is not always our intent that the notional worlds we construct within our notional worlds be *fiction*. For instance, coming upon

Smith's corpse, we might reconstruct the crime in our imagination—a different way of thinking about Smith's murderer.

That ignoring these distinctions has contributed to the
confusion in the discussions of *de re* and *de dicto* is clear, I trust.
Just consider the infamous case of believing that the shortest spy
is a spy. It is commonly presumed that we all know what state of
mind is being alluded to in this example, but in fact there are
many quite different possibilities undistinguished in the literature. Does Tom believe that the shortest spy is a spy in virtue of
nothing more than a normal share of logical acumen, or does he
also have to have a certain sort of idleness and a penchant for
reflecting on tautologies? (Should we say we all believe this, and
also that the tallest tree is a tree, and so forth ad infinitum?) Does
Tom believe that the shortest dugong is a dugong? Or, to pursue
a different tack, what is the relation between believing that that
man is a spy and thinking that that man is a spy? Sincere denial
is often invoked as a telling sign of disbelief. What is it best
viewed as showing? And so forth. When good psychological
accounts of these notional attitude phenomena are in hand, some
of the puzzles about reference will be solved and others will be
discredited. I doubt if there will be any residue.[44]

Notes to chapter 1

[1] See Burge (1979a) for novel arguments indirectly supporting the claim that the
ordinary notion of belief individuates beliefs propositionally.

[2] Field (in a postscript to Field 1978, to appear in Block forthcoming) sees this view
leading inevitably to the strong claim:

> 'The theory of measurement ... explains why real numbers can be used to
> "measure" mass (better: to serve as a scale for mass). It does this in the following way.
> First, certain properties and relations among massive objects are cited—properties and
> relations that are specifiable without reference to numbers. Then a representation
> theorem is proved: such a theorem says that if any system of objects has the properties
> and relations cited, then there is a mapping of that system into the real numbers which
> "preserves structure". Consequently, assigning real numbers to the objects is a
> convenient way of discussing the intrinsic mass-relations that those objects have, but
> those intrinsic relations don't themselves require the existence of real numbers. ...
> Can we solve Brentano's problem of the intentionality of propositional attitudes in
> an analogous way? To do so we would have to postulate a *system of entities* [my italics]
> inside the believer which was related via a structure-preserving mapping to the system
> of propositions. The "structure" that such a mapping would have to preserve would be
> the kind of structure important to propositions: viz., logical structure, and this I think

means that the system of entities inside the believer can be viewed as a system of sentences—an internal system of representation.'

For massive objects we postulate or isolate 'properties and relations'; why not properties and relations, rather than 'a system of entities', in the case of psychological subjects? Whatever might be 'measured' by propositional attitude predicates is measured indirectly, and Field's opting for a language of thought to 'explain' the success (the extent of which is still uncharted) of propositional measurement is a premature guess, not an implication of this view of the predicates.

[3] *Having a big red nose* is a state that can figure prominently in one's psychology, but it is not itself a psychological state. *Believing that one has a big red nose* is one of the many psychological states that would typically go with having a big red nose, and without which the state of having a big red nose would tend to be psychologically inert (like having a big red liver). (This just ostends, and does not pretend to explicate, the intuitive distinction between psychological states and the other states of a creature.)

[4] One of the simplest and most convincing is due to Vendler (in conversation): suppose over a ten year period I believe that Angola is an independent nation. Intuitively, this is a *constancy* of psychological state—something about me that does not change—and yet this one belief of mine can change in truth-value during the decade. Counting my state as one enduring belief, it cannot be a *propositional* attitude.

[5] One could sum up the case Putnam, Kaplan and Perry present thus: propositions are not *graspable* because they can *elude* us; the presence or absence of a particular proposition 'in our grasp' can be psychologically irrelevant.

[6] The issues running in the mental-words-versus-mental-pictures controversies are largely orthogonal to the issues discussed here, which concern problems that must be solved before *either* mental images or mental sentences could be given a clean bill of health as theoretical entities.

[7] John McCarthy claims this is too strong: the purely formal patterns of repetition and co-occurrence to be found in book-length strings of characters put a very strong constraint—which might be called the cryptographer's constraint—on anyone trying to devise non-trivially different interpretations of a text. A number of 'cheap tricks' will produce different interpretations of little interest. For instance, declare first person singular in English to be a variety of third-person singular in Schmenglish, and (with a little fussing), turn an autobiography into a biography. Or declare Schmenglish to have very long words—English chapter length, in fact—and turn any ten-chapter book into the ten-word sentence of your choice. The prospect of interestingly different interpretations of a text is hard to assess, but worth exploring, since it provides a boundary condition for 'radical translation' (and hence 'radical interpretation'—see Lewis 1974) thought-experiments.

[8] Field (1978) notices this problem, but, astonishingly, dismisses it; '... the notion of type-identity between tokens in one organism and tokens in the other is not needed for psychological theory, and can be regarded as a meaningless notion' (p. 58, note 34). His reasons for maintaining this remarkable view are no less remarkable—but too devious to do justice to explicitly here. There are many points of agreement and disagreement of importance between Field's paper and this one beyond those I shall discuss, but a discussion of them all would double the length of this paper. I discuss Fodor's (1975) commitment to sententialism in Dennett (1977) (reprinted in Dennett 1978a).

[9] Cf. also Field (1978, p. 47), who considers such claims as 'He believes some sentence of his language which plays approximately the role in his psychology that the sentence "there's a rabbit nearby" plays in mine'. He decides that such claims involve introducing a 'more-or-less semantic notion' into a psychological theory that was supposed to be

liberated from semantic problems. Cf. also Stich (this volume) on content ascription and content similarity.

[10] Fodor (1980) makes much the same point in arguing for what he calls the formality condition: mental states can be (type) distinct only if the representations which constitute their objects are formally distinct. See also Field (1978).

[11] A 'direct' test is still not foolproof, of course, or *absolutely* direct.

[12] I am using 'links' as a wild-card for whatever is needed to play this role; no one yet knows (so far as I know) how to solve this problem in detail.

[13] Charles Taylor's discussions of explicitness and 'explicitation' helped shape the claims in the rest of this section.

[14] 'Subserves' is a useful hand-waving term for which we may thank the neuro-physiologists. Putting two bits of jargon together, we can say a belief *supervenes* on the state that *subserves* it.

[15] Burge (1979a) presents an extended thought-experiment about beliefs about arthritis that can be seen as drawing the boundary between the system proper and its environment *outside* the biological individual entirely; the contextual variations involve *social* practices outside the experience of the subject.

[16] On some of the vicissitudes of auto-phenomenology, see my 'Two Approaches to Mental Images' in Dennett (1978a). See also Campbell (forthcoming), on reasons for doing the epistemology of the other.

[17] What about the objects of its fears, hopes and desires? Are they denizens of the subject's notional world, or must we add a desire-world, a fear-world, and so forth to the subject's belief-world? (Joe Camp and others have pressed this worry.) When something the subject believes to exist is also feared, or desired, by him, there is no problem: some denizen of his notional world is simply coloured with desire or fear or admiration or whatever. How to treat 'the dream house I hope someday to build' is another matter. Postponing the details to another occasion, I will venture some incautious general claims. My dream house is not a denizen of my notional world on a par with my house or even the house I will end my days in; thinking about *it* (my dream house) is not, for instance, to be analysed in the same fashion as thinking about my house or thinking about the house I will end my days in. (More on this theme in the following section.) My dream house gets constituted indirectly in my notional world via what we might call my *specifications*, which are perfectly ordinary denizens of my notional world, and my general beliefs and other attitudes. I believe in my specifications, which already exist in the world as items of mental furniture created by my thinking, and then there are general beliefs and desires and the like involving those specifications: to say my dream house is built of cedar is not to say my specification is made of cedar, but to say that any house built to my specification would be of made cedar. To say I plan to build *it* next year is to say that I plan to build a house to my specs next year.

[18] Special features of (literary) fiction lead Lewis to make substantial ingenious modifications to this idea, in order to account for the role of background assumptions, narrator knowledge, and the like, in the normal interpretation of fiction. For instance, we assume that the map of Holmes's London is that of Victorian London except where overridden by Conan Doyle's inventions; the texts neither assert nor strictly imply that Holmes did not have a third nostril, but the possible worlds in which this is the case are excluded.

[19] Variations on the idea of treating a 'belief-world' as the set of all possible worlds in which all one's beliefs are true have been given much discussion since the idea was introduced by Hintikka (1962). Sorting out the variations and comparing and contrasting

them with my development of the theme is some scholarship to be saved for another occasion.

[20] The issues surrounding 'I' and indexicality are much more complicated than this hurried acknowledgement reveals. See not only Perry and Lewis, but also Castañeda (1966, 1967, 1968). For illuminating reflections on a similar theme, see Hofstadter (1979), pp. 373–6.

[21] 'A man may think that he believes p, while his behaviour can only be explained by the hypothesis that he believes not-p, given that it is known that he wants z. Perhaps the confusion in his mind cannot be conveyed by any simple [or complex—D.C.D.] account of what he believes: perhaps only a reproduction of the complexity and confusion will be accurate.' (Hampshire 1975, p. 123.)

[22] Cf. the discussion of Phenomenology and 'Feenomanology' in 'Two Approaches to Mental Images', in Dennett (1978a). See also Lewis's (1978) remarks on how to deal with inconsistency in a work of fiction.

[23] Putnam (1975a) describes the theory of meaning available to methodological solipsism as a theory of 'individual competence' (p. 246).

[24] Kaplan (1968) is explicit: 'The crucial feature of this notion [Ralph's vivid names] is that it depends only on Ralph's current mental state, and ignores all links whether by resemblance or genesis with the actual world. ... It is intended to go to the purely internal aspects of individuation.' (p. 201.)

[25] My *Doppelgänger* would not, however, have thoughts *about me* when he thought 'I'm sleepy' and so forth. The reference of the first person pronoun is not affected by world-switching, of course (see Putnam 1975a; Perry 1977, 1979; Lewis 1979). But one must be careful not to inflate this point into a metaphysical doctrine about personal identity. Consider this variation on a familiar science-fiction theme in philosophy. Your space ship crashes on Mars, and you want to return to Earth. Fortunately, a Teleporter is available. You step into the booth on Mars and it does a complete microphysical analysis of you, which requires dissolving you into your component atoms, of course. It beams the information to Earth, where the receiver, stocked with lots of atoms the way a photocopier is stocked with fresh white paper, creates an exact duplicate of you, which steps out and 'continues' your life on Earth with your family and friends. Does the Teleporter 'murder to dissect', or has it transported you home? When the newly arrived Earth-you says 'I had a nasty accident on Mars', is what he says true? Suppose the Teleporter can obtain its information about you without dissolving you, so that you continue a solitary life on Mars. On your mark, get set, go. . .

[26] Connoisseurs of the literature will note that this sentence delicately reproduces a familiar equivocation that marks the *genre*: does '*de re*' modify 'speaking' or 'beliefs'; does '*de dicto*' modify 'attributions' or 'belief'?

[27] Evans, in lectures in Oxford in 1979, developed the theme of the *process* of keeping track of things in the world. It echoes a central theme in Neisser's (1976) apostatic renunciation of two-dimensional, tachistoscopic experiments in the psychology of perception, in favour of a Gibsonian 'ecological' approach to perception.

[28] The problem of preserving that match has as its core the 'frame problem' of Artificial Intelligence which arises for planning systems that must reason about the effects of contemplated actions. See McCarthy and Hayes (1969), and Dennett (1978a), ch. 7. It is either the most difficult problem A.I. must—and can eventually—solve, or the *reductio ad absurdum* of mental representation theory.

[29] 'One may accept the Brentano thesis either as showing the indispensability of intentional idioms and the importance of an autonomous science of intention, or as

showing the baselessness of intentional idioms and the emptiness of a science of intention. My attitude, unlike Brentano's, is the second.' (Quine 1960, p. 221.)

[30] Quine (1969) contrasts this 'portentous' belief with 'trivial' beliefs, such as the belief that the shortest spy is a spy—but later in the same piece is led to a view that 'virtually annuls the seemingly vital contrast between [such beliefs]. ... At first this seems intolerable, but it grows on one.' It does, with suitable cultivation, which I shall try to provide.

[31] Quine explicitly renounces this course, while still hoping to capture whatever useful, important inferences there are. (1960, p. 221; 1969.) For further arguments in favour of viewing some cases of psychological states as *de re* attitudes toward properties and relations, see Aquila (1977), especially pp. 84–92.

[32] See, e.g., 'What do General Propositions Refer to?' and 'Oratio Obliqua' in Prior (1976).

[33] Kripke's views on unicorns require that not only are there no unicorns, but there could not be any. Could there then be the attribute of unicornhood? See the Addendum to Kripke (1972) (pp. 763–9).

[34] It would thus also be true that *you* are talking *about her* (in this weak sense) when you assert that I was thinking about the shortest spy. Thus I must disagree with Kripke: 'If a description is embedded in a *de dicto* (intensional) context, we cannot be said to be talking *about* the thing described, either *qua* its satisfaction of the description or *qua* anything else. Taken *de dicto*, "Jones believes that the richest debutante in Dubuque will marry him", can be asserted by someone who thinks (let us suppose, wrongly) that there are no debutantes in Dubuque; certainly then, he is *in no way* [my italics] talking about the richest debutante, even "attributively"' (1977). Imagine Debby, who knows quite well she is the richest debutante in Dubuque, overhearing this remark; she might comment: 'You may not know it, buster, but you're talking about me, and you've given me a good laugh. For though Jones doesn't know me, I know who he is—the social climbing little jerk—and he hasn't a prayer.'

[35] Sellars (1974) discusses the formula: 'Jones believes with respect to someone (who may or may not be real) that he is wise'.

[36] In this regard mental sayings are seen to be like *qualia*: *apparently* intrinsic features of minds to be contrasted with relational, functionalistically characterizable features. I argue for the incoherence of this vision of qualia in 'Quining Qualia' (unpublished). For a clear expression of the suspect view, consider Burge (1977), p. 345: 'From a semantical viewpoint, a *de dicto* belief is a belief in which the believer is related only to a completely expressed proposition (*dictum*).'

[37] Kaplan (forthcoming) makes disarmingly explicit use of this image in proposing to 'use the distinction between direct and indirect discourse to match the distinction between character and content' (of thoughts or beliefs, now, not sentences); this, in spite of his equally disarming concession that 'there is no real syntax to the language of thought'.

[38] When Burge (1977) doubts the existence of any 'pure' *de dicto* beliefs he is best read, I think, as expressing an appreciation of this point. See also Field (1978), pp. 21–3.

[39] House (unpublished) provides an account of some of the principles governing such derivations.

[40] Hornsby (1977) discusses the case of 'Jones, an uneducated individual, who has ... found "Quine" on a list of names of philosophers. He knows nothing whatsoever about that man, but comes simply to believe that one could assert something true with the words "Quine is a philosopher". In such circumstances as these a relational reading ... cannot be right. But in these circumstances Jones really believes no more than that the properties of being called "Quine" and being a philosopher are somewhere coinstantiated.'

(p. 47.) The last sentence is *a* relational reading, however, and depending on circumstances further relational readings are likely: e.g., the property of being the person named 'Quine' whom the authors of this list intended to include on the list is uniquely instantiated by someone who also instantiates the property of being a philosopher.

[41] Richmond Thomason, in conversation, suggests that the problems raised by these cases (by (1), (2), (25) and (29)) are actually problems about the logic of plurals. He would distinguish believing that dugongs are mammals from believing that all dugongs are mammals. Perhaps we must make this distinction—and if so, it's a pity, since Thomason claims no one has yet devised a trouble-free account of plurals—but even if we make the distinction, I think nothing but naturalness is lost if we recast my examples explicitly as cases of believing all snakes are slimy, all whales are mammals and so forth.

[42] Not all believers in the causal theory carve up the cases the same way. Vendler (forthcoming) would insist that even in the case where I do not see the dropping of the gum wrapper, *since only one person could have dropped that gum wrapper*—since only one person could have made that Kripkean 'insertion into history'—my belief is rigidly and strongly *about* that person. 'Are we not acquainted, in a very real sense, with the otherwise unknown slave who left his footprint in King Tut's tomb? Or with the scribe who carved *that* particular hieroglyph into *that* stone four thousand years ago?' I suspect that Vendler's apparently extreme position is the only stable position for a causal theorist to adopt. But perhaps I have misinterpreted Vendler; perhaps *someone else* could have dropped *that* gum wrapper, but *no one else* could have left *that* footprint. Then Tom's belief about whoever made the footprint in the wet concrete, to the effect that he is a careless slob, is directly about *that* individual in a way his belief about the dropper of the gum wrapper is not directly about *him*. Not for nothing have several wits called the causal theory the Doctrine of Original *Sinn*.

[43] Consider the contrast between

(a) I enter a telephone booth and find a dime in the coin return box; I believe that whoever used the booth last left a dime in the box.

(b) I enter a telephone booth, make a call and deliberately leave a dime in the coin box; I believe that whoever uses the booth next will find a dime in the coin return box.

Is my belief in (b) *already* about some particular person? But I haven't the faintest idea who that is or will be! (Cf. Harman 1977.) So what? (Cf. Searle 1979.) I don't have the faintest idea whom my belief in (a) is about either, and in fact it is much likelier that I can discover the object of my belief in (b) than in (a). If I believe whoever left the dime kidnapped Jones, the kidnapper is probably quite safe; if I believe whoever finds the dime is the kidnapper (it's a signal in the ransom exchange scheme) capture is more likely.

[44] I am grateful to all the patient souls who have tried to help me see my way out of this project. In addition to all those mentioned in the other notes and bibliography, I want to acknowledge the help of Peter Alexander, David Hirschmann, Christopher Peacocke, Pat Hayes, John Haugeland, Robert Moore, Zenon Pylyshyn, Paul Benacerraf, and Dagfinn Føllesdal. This research was supported by an N.E.H. Fellowship, and by the National Science Foundation (BNS 78–24671) and the Alfred P. Sloan Foundation.

2

Other Bodies

Tyler Burge

It is fairly uncontroversial, I think, that we can conceive a person's behaviour and behavioural dispositions, his physical acts and states, his qualitative feels and fields (all non-intentionally described) as remaining fixed, while his mental attitudes of a certain kind—his *de re* attitudes—vary.[1] Thus we can imagine Alfred's believing of apple 1 that it is wholesome, and holding a true belief. Without altering Alfred's dispositions, subjective experiences, and so forth, we can imagine having substituted an identically appearing but internally rotten apple 2. In such a case, Alfred's belief differs while his behavioural dispositions, inner causal states and qualitative experiences remain constant.

This sort of point is important for understanding mentalistic notions and their role in our cognitive lives. But taken by itself, it tells us nothing very interesting about mental states. For it is easy (and I think appropriate) to phrase the point so as to strip it of immediate philosophical excitement. We may say that Alfred has the *same* belief-content in both situations.[2] It is just that he would be making contextually different applications of that content to different entities. His belief is true of apple 1 and false of apple 2. The *nature* of his mental state is the same. He simply bears different relations to his environment. We do say in ordinary language that one belief is true and the other is false. But this is just another way of saying that what he believes is true of the first apple and would be false of the second. We may call these relational beliefs different beliefs if we want. But differences among such relational beliefs do not entail differences among mental states or contents, as these have traditionally been viewed.

This deflationary interpretation seems to me to be correct. But it suggests an oversimplified picture of the relation between a person's mental states or events and public or external objects and events. It suggests that it is possible to separate rather neatly what aspects of propositional attitudes depend on the person holding the attitudes and what aspects derive from matters external. There is no difference in the obliquely occurring expressions in the content-clauses we attribute to Alfred. It is these sorts of expressions that carry the load in characterizing the individual's mental states and processes. So it might be thought that we could explicate such states and processes by training our philosophical attention purely on the individual subject, explicating the differences in the physical objects that his content applies to in terms of facts about his environment.

To present the view from a different angle: *de re* belief-attributions are fundamentally predicational. They consist in applying or relating an incompletely interpreted content-clause, an open sentence, to an object or sequence of objects, which in effect completes the interpretation. What objects these open sentences apply to may vary with context. But, according to the picture, it remains possible to divide off contextual or environmental elements represented in the propositional attitude attributions from more specifically mentalistic elements. Only the constant features of the predication represent the latter. And the specifically mental features of the propositional attitude can, according to this picture, be understood purely in *individualistic* terms—in terms of the subject's internal acts and skills, his internal causal and functional relations, his surface stimulations, his behaviour and behavioural dispositions, and his qualitative experiences, all non-intentionally characterized and specified without regard to the nature of his social or physical environment.

The aim of this paper is to bring out part of what is wrong with this picture and to make some suggestions about the complex relation between a person's mental states and his environment. Through a discussion of certain elements of Putnam's Twin-Earth examples, I shall try to characterize ways in which identifying a person's mental states depends on the nature of his physical environment—or on specification, by his fellows, of the nature of that environment.[3]

Before entering into the details of Putnam's thought-experiment, I want to sketch the general position that I shall be defending. What is right and what is wrong in the viewpoint I set out in the third and fourth paragraphs of this paper? I have already given some indication of what seems right: individual entities referred to by transparently occurring expressions, and, more generally, entities (however referred to or characterized) *of* which a person holds his beliefs do not in general play a direct role in characterizing the nature of the person's mental state or event. The difference between apples 1 and 2 does not bear on Alfred's mind in any sense that would immediately affect explanation of Alfred's behaviour or assessment of the rationality of his mental activity. Identities of and differences among physical objects are crucial to these enterprises only in so far as those identities and differences affect Alfred's way of viewing such objects.[4] Moreover, it seems unexceptionable to claim that the obliquely occurring expressions in propositional attitude attributions are critical for characterizing a given person's mental state. Such occurrences are the stuff of which explanations of his actions and assessments of his rationality are made.

What I reject is the view that mental states and processes individuated by such obliquely occurring expressions can be understood (or 'accounted for') purely in terms of non-intentional characterizations of the *individual subject's* acts, skills, dispositions, physical states, 'functional states', and the effects of environmental stimuli on him, without regard to the nature of his physical environment or the activities of his fellows.[5]

In 'Individualism and the Mental' (Burge 1979a) I presented a thought-experiment in which one fixed non-intentional, individualistic descriptions of the physical, behavioural, phenomenalistic, and (on most formulations) functional histories of an individual. By varying the *uses of words* in his linguistic community we found that the contents of his propositional attitudes varied.[6] I shall draw a parallel conclusion from Putnam's Twin-Earth thought-experiment: We can fix an individual's physical, behavioural, phenomenalistic, and (on some formulations) functional histories; by varying the *physical environment* one finds that the contents of his propositional attitudes vary. It is to be re-emphasized that the variations in propositional attitudes envisaged are not exhausted by variations in the entities to which individuals' mental contents are related. The contents themselves vary. At any rate, I shall so argue.

I

In Putnam's thought-experiment, we are to conceive of a near duplicate of our planet Earth, called 'Twin-Earth'. Except for certain features about to be noted (and necessary consequences of these features), Twin-Earth duplicates Earth in every detail. The physical environments look and largely are the same. Many of the inhabitants of one planet have duplicate counterparts on the other, with duplicate macro-physical, experiential and dispositional histories.

One key difference between the two planets is that the liquid on Twin-Earth that runs in rivers and faucets, and is called 'water' by those who speak what is called 'English' is not H_2O, but a different liquid with a radically different chemical formula, XYZ. I think it natural and obviously correct to say, with Putnam, that the stuff that runs in rivers and faucets on Twin-Earth is thus not water. I shall not argue for this view because it is pretty obvious, pretty widely shared, and stronger than arguments that might be or have been brought to buttress it. I will just assume that XYZ is not water of any sort. Water is H_2O. What the Twin-Earthians call 'water' is XYZ. In translating into English occurrences of 'water' in the mouths of Twin-Earthians, we would do best to coin a new non-scientific word (say, 'twater'), explicated as applying to stuff that looks and tastes like water, but with a fundamentally different chemistry.

It is worth bearing in mind that the thought-experiment might apply to any relatively non-theoretical natural kind word. One need not choose an expression as central to our everyday lives as 'water' is. For example, we could (as Putnam in effect suggests) imagine the relevant difference between Earth and Twin-Earth to involve the application of 'aluminium', or 'elm', or 'mackerel'.[7]

A second key difference between Earth and Twin-Earth—as we shall discuss the case—is that the scientific community on Earth has determined that the chemical structure of water is H_2O, whereas the scientific community on Twin-Earth knows that the structure of twater is XYZ. These pieces of knowledge have spread into the respective lay communities, but have not saturated them. In particular, there are numerous scattered individuals on Earth and Twin-Earth untouched by the scientific developments. It is these latter individuals who have duplicate counterparts.

We now suppose that Adam is an English speaker and that Adam$_{te}$ is his counterpart on Twin-Earth. Neither knows the chemical properties of what he calls 'water'. This gap in their knowledge is probably not atypical of uneducated members of their communities.[8] But similar gaps are clearly to be expected of users of terms like 'aluminium', 'elm', 'mackerel'. (Perhaps not in the case of water, but in the other cases, we could even imagine that Adam and Adam$_{te}$ have no clear idea of what the relevant entities in their respective environments look like or how they feel, smell, or taste.) We further suppose that both have the same qualitative perceptual intake and qualitative streams of consciousness, the same movements, the same behavioural dispositions and inner functional states (non-intentionally and individualistically described). In so far as they do not ingest, say, aluminium or its counterpart, we might even fix their physical states as identical.

When Adam says or consciously thinks the words, 'There is some water within twenty miles, I hope', Adam$_{te}$ says or consciously thinks the same word forms. But there are differences. As Putnam in effect points out, Adam's occurrences of 'water' apply to water and mean *water*, whereas Adam$_{te}$'s apply to twater and mean *twater*. And, as Putnam does not note, the differences affect *oblique* occurrences in 'that'-clauses that provide the contents of their mental states and events. Adam hopes that there is some water (oblique occurrence) within twenty miles. Adam$_{te}$ hopes that there is some twater within twenty miles. That is, even as we suppose that 'water' and 'twater' are not logically exchangeable with co-extensive expressions *salva veritate*, we have a difference between their thoughts (thought contents).

Laying aside the indexical implicit in 'within twenty miles', the propositional attitudes involved are not even *de re*. But I need not argue this point. Someone might wish to claim that these are *de re* attitudes about the relevant *properties*—of water (being

water? waterhood?) in one case and of twater (etc.) in the other.
I need not dispute this claim here. It is enough to note that even
if the relevant sentences relate Adam and his counterpart to *res*,
those sentences also specify *how* Adam and Adam$_{te}$ think about
the *res*. In the sentence applied to Adam, 'water' is, by hypothesis,
not exchangeable with co-extensive expressions. It is not exchangeable with 'H_2O', or with 'liquid which covers two-thirds
of the face of the earth', or with 'liquid said by the Bible to flow
from a rock when Moses struck it with a rod'. 'Water' occurs
obliquely in the relevant attribution. And it is expressions in
oblique occurrence that play the role of specifying a person's
mental contents, what his thoughts are.

In sum, mental states and events like beliefs and thoughts are
individuated partly by reference to the constant, or obliquely
occurring, elements in content clauses. But the contents of Adam
and Adam$_{te}$'s beliefs and thoughts differ while every feature of
their non-intentionally and individualistically described physical, behavioural, dispositional, and phenomenal histories remains
the same. Exact identity of physical states is implausible in the
case of water. But this point is irrelevant to the force of the
example—and could be circumvented by using a word, such as
'aluminium', 'elm', etc., that does not apply to something Adam
ingests. The difference in their mental states and events seems to
be a product primarily of differences in their physical environments, mediated by differences in their social environments—in
the mental states of their fellows and conventional meanings of
words they and their fellows employ.

II

The preceding argument and its conclusion are not to be found
in Putnam's paper. Indeed, the conclusion appears to be
incompatible with some of what he says. For example, Putnam
interprets the difference between Earth and Twin-Earth uses of
'water' purely as a difference in extension. And he states that the
relevant Earthian and Twin-Earthian are 'exact duplicates in
... feelings, thoughts, interior monologue etc.' (1975b, p. 224).
On our version of the argument the two are in no sense exact
duplicates in their thoughts. The differences show up in oblique

occurrences in true attributions of propositional attitudes. I shall not speculate about why Putnam did not draw a conclusion so close to the source of his main argument. Instead, I will criticize aspects of his discussion that tend to obscure the conclusion (and have certainly done so for others).

Chief among these aspects is the claim that natural kind words like 'water' are indexical (ibid., pp. 229–35). This view tends to suggest that Earth and Twin-Earth occurrences of 'water' can be assimilated simply to occurrences of indexical expressions like 'this' or 'I'. Adam and Adam$_{te}$'s propositional attitudes would then be further examples of the kind of *de re* attitudes mentioned at the outset of this paper. Their contents would be the same, but would be applied to different *res*.[9] If this were so, it might appear that there would remain a convenient and natural means of segregating those features of propositional attitudes that derive from the nature of a person's social and physical context, on one hand, from those features that derive from the organism's nature, and palpable effects of the environment on it, on the other. The trouble is that there is no appropriate sense in which natural kind terms like 'water' are indexical.

Putnam gives the customary explication of the notion of indexicality: 'Words like "now", "this" "here", have long been recognized to be *indexical* or *token-reflexive*—i.e. to have an extension which varied from context to context or token to token' (ibid., pp. 233–4). I think that it is clear that 'water', *interpreted as it is in English*, or as we English speakers standardly interpret it, does not shift extension from context to context in this way. (One must, of course, hold the language, or linguistic construal, fixed. Otherwise, every word will trivially count as indexical. For by the very conventionality of language, we can always imagine some context in which our word—word form—has a different extension.) The extension of 'water', as interpreted in English in all non-oblique contexts, is (roughly) the set of all aggregates of H_2O molecules, together, probably, with the individual molecules. There is nothing at all indexical about 'water' in the customary sense of 'indexical'.

Putnam suggests several grounds for calling natural kind

words indexical. I shall not try to criticize all of these, but will
deal with a sampling:

(a) Now then, we have maintained that indexicality extends beyond
the *obviously* indexical words . . . Our theory can be summarized
as saying that words like 'water' have an unnoticed indexical
component: 'water' is stuff that bears a certain similarity relation
to the water *around here*. Water at another time or in another
place or even in another possible world has to bear the relation
[same-liquid] to *our* 'water' in order to be water. [1975b, p. 234.]

(b) 'Water' is indexical. What do I mean by that? If it is indexical,
if what I am saying is right, then 'water' means 'whatever is like
water, bears some equivalence relation, say the liquid relation,
to *our* water.' Where 'our' is, of course, an indexical word. If
that's how the extension of 'water' is determined, then the
environment determines the extension of 'water'. Whether 'our'
water is in fact XYZ or H_2O.[10]

These remarks are hard to interpret with any exactness because
of the prima facie circularity, or perhaps ellipticality, of the
explications. Water around here, or our water, is just water.
Nobody else's water, and no water anywhere else, is any different.
Water is simply H_2O (give or take some isotopes and impurities).
These points show the superfluousness of the indexical expressions. No shift of extension with shift in context needs to be
provided for.

Narrower consideration of these 'meaning explanations' of
'water' brings out the same point. One might extrapolate from
(a) the notion that 'water' *means* (a') 'stuff that bears the same-
liquid relation to the stuff we call "water" around here'. But this
cannot be right. (I pass over the fact that there is no reason to
believe that the meaning of 'water' involves reference to the
linguistic expression 'water'. Such reference could be eliminated.)
For if Adam and his colleagues visited Twin-Earth and (still
speaking English) called XYZ 'water', it would follow on this
meaning explication that their uses of the sentence 'Water flows
in that stream' would be true. They would make no mistake in
speaking English and calling XYZ 'water'. For since the
extension of 'here' would shift, occurrences on Twin-Earth of
'stuff that bears the same-liquid relation to the stuff we call

"water" around here flows in that stream' would be true. But by
Putnam's own account, which is clearly right on this point, there
is no water on Twin-Earth. And there is no reason why an
English speaker should not be held to this account when he visits
Twin-Earth. The problem is that although 'here' shifts its
extension with context, 'water' does not. 'Water' lacks the
indexicality of 'here'.

A similar objection would apply to extrapolating from (b) the
notion that 'water' means (b') 'whatever bears the same-liquid
relation to what we call "water"', or (b″) 'whatever bears the
same-liquid relation to this stuff'. 'Water' *interpreted as it is in
English* does not shift its extension with shifts of speakers, as (b')
and (b″) do. The fact that the Twin-Earthians apply 'water' to
XYZ is not a reflection of a shift in extension of an indexical
expression with a fixed linguistic (English) meaning, but of a
shift in meaning between one language, and linguistic community, and another. Any expression, indexical or not, can undergo
such 'shifts', as a mere consequence of the conventionality of
language. The relevant meaning equivalence to (b') is no more
plausible than saying that 'bachelor' is indexical because it means
'whatever social role the speaker applies "bachelor" to' where
'the speaker' is allowed to shift in its application to speakers of
different linguistic communities according to context. If Indians
applied 'bachelor' to all and only male hogs, it would not follow
that 'bachelor' as it is used in English is indexical. Similar points
apply to (b″).

At best, the term 'water' and a given occurrence or token of
(b') or (a'), say an introducing token, have some sort of deep or
necessary equivalence. But there is no reason to conclude that the
indexicality of (a'), (b') or (b″)—which is a feature governing
general use, not particular occurrences—infects the meaning of
the *expression* 'water', as it is used in English.

Much of what Putnam says suggests that the appeal to
indexicality is supposed to serve other desiderata. One is a desire
to defend a certain view of the role of the natural kind terms in
talk about necessity. Roughly, the idea is that 'water' applies to
water in all discourse about necessity. Putnam expresses this idea
by calling natural kind terms rigid, and seems to equate

indexicality and rigidity (1975b, p. 234). These points raise a morass of complex issues which I want to avoid getting into. It is enough here to point out that a term can be rigid without being indexical. Structural descriptive syntactical names are examples. Denying that natural kind terms are indexical is fully compatible with holding that they play any given role in discourse about necessity.

Another purpose that the appeal to indexicality seems to serve is that of accounting for the way natural kind terms are introduced, or the way their reference is fixed (cf. 1975b, p. 234, and (b) above). It may well be that indexicals frequently play a part in the (reconstructed) introduction or reference-fixing of natural kind terms. But this clearly does not imply that the natural kind terms partake in the indexicality of their introducers or reference-fixers. With some stage setting, one could introduce natural kind terms, with all their putative modal properties, by using non-indexical terms. Thus a more general rational reconstruction of the introduction, reference-fixing, and modal behaviour of natural kind terms is needed. The claim that natural kind terms are themselves indexical is neither a needed nor a plausible account of these matters.

It does seem to me that there is a grain of truth encased within the claim that natural kind terms are indexical. It is this. *De re* beliefs usually enter into the reference-fixing of natural kind terms. The application of such terms seems to be typically fixed partly by *de re* beliefs we have of particular individuals, or quantities of stuff, or physical magnitudes or properties—beliefs that establish a semantical relation between term and object. (Sometimes the *de re* beliefs are about evidence that the terms are introduced to explain.) Having such beliefs requires that one be in not-purely-context-free conceptual relations to the relevant entities. (Cf. Burge 1977.) That is, one must be in the sort of relation to the entities that someone who indexically refers to them would be. One can grant the role of such beliefs in establishing the application and function of natural kind terms, without granting that all beliefs and statements involving terms whose use is so established are indexical. There seems to be no justification for the latter view, and clear evidence against it.

I have belaboured this criticism not because I think that the claim about indexicality is crucial to Putnam's primary aims in 'The Meaning of "Meaning"'. Rather, my purpose has been to clear an obstacle to properly evaluating the importance of the Twin-Earth example for a philosophical understanding of belief and thought. The difference between mistaking natural kind words for indexicals and not doing so—rather a small linguistic point in itself—has large implications for our understanding of mentalistic notions. Simply assimilating the Twin-Earth example to the example of indexical attitudes I gave at the outset trivializes its bearing on philosophical understanding of the mental. Seen aright, the example suggests a picture in which the individuation of a given individual's mental *contents* depends partly on the nature (or what his fellows think to be the nature) of entities about which he or his fellows have *de re* beliefs. The identity of one's mental contents, states, and events is not independent of the nature of one's physical and social environment.

To summarize our view: The differences between Earth and Twin-Earth will affect the attributions of propositional attitudes to inhabitants of the two planets, including Adam and Adam$_{te}$. The differences are not to be assimilated to differences in the extensions of indexical expressions with the same constant, linguistic meaning. For the relevant terms are not indexical. The differences, rather, involve the constant context-free interpretation of the terms. Propositional attitude attributions which put the terms in oblique occurrence will thus affect the content of the propositional attitudes. Since mental acts and states are individuated (partly) in terms of their contents, the differences between Earth and Twin-Earth include differences in the mental acts and states of their inhabitants.

III

Let us step back now and scrutinize Putnam's interpretation of his thought-experiment in the light of the fact that natural kind terms are not indexical. Putnam's primary thesis is that a person's psychological states—in what Putnam calls the 'narrow sense' of this expression—do not 'fix' the extensions of the terms the person uses. A psychological state in the 'narrow sense' is said to be one which does not 'presuppose' the existence of any individual other than the person who is in that state (1975b, p. 220). The term 'presuppose' is, of course, notoriously open to a variety of uses. But Putnam's glosses seem to indicate that a person's being in a psychological state does not presuppose a proposition P, if it does not logically entail P.[11]

Now we are in a position to explore a first guess about what psychological states are such in the 'narrow sense'.[12] According to this interpretation, being in a psychological state in the narrow sense (at least as far as propositional attitudes are concerned) is to be in a state correctly ascribable in terms of a content-clause which contains no expressions in a position (in the surface grammar) which admits of existential generalization, and which is not in any sense *de re*. *De dicto*, non-relational propositional attitudes would thus be psychological states in the narrow sense. They entail by existential generalization the existence of no entities other than the subject (and his thought contents). *De re* propositional attitudes—at least those *de re* propositional attitudes in which the subject is characterized as being in relation to some thing other than himself and his thought contents—appear to be psychological states in the 'wide sense'. Having *de re* attitudes of (*de*) objects other than oneself entails the existence of objects other than oneself.

Granted this provisional interpretation, the question arises whether Putnam's Twin-Earth examples show that a person's psychological states in the narrow sense fail to 'fix' the extensions of the terms he uses. It would seem that to show this, the examples would have to be interpreted in such a way that Adam and Adam$_{te}$ would have the same *de dicto* propositional attitudes while the extensions of their terms differed. This objective would suggest an even stronger interpretation of the thought-experiment. Expressions in oblique position in true attributions of attitudes to Adam and Adam$_{te}$ would be held constant while the extensions of their terms varied. But neither of these interpretations is plausible.[13]

Let us see why. To begin with, it is clear that Adam and Adam$_{te}$ will (or might) have numerous propositional attitudes correctly attributable with the relevant natural kind terms in oblique position. The point of such attributions is to characterize a subject's mental states and events in such a way as to take into account the *way* he views or thinks about objects in his environment. We thus describe his perspective on his environment and utilize such descriptions in predicting, explaining, and assessing the rationality and the correctness or success of his mental processes and overt acts. These enterprises of explanation and assessment provide much of the point of attributing propositional attitudes. And the way a subject thinks about natural kinds of stuffs and things is, of course, as relevant to these enterprises as the way he thinks about anything else. Moreover, there is no intuitive reason to doubt that the relevant natural kind terms can express and characterize his way of thinking about the relevant stuffs and things—water, aluminium, elms, mackerel. The relevant subjects meet socially accepted standards for using the terms. At worst, they lack a specialist's knowledge about the structure of the stuffs and things to which their terms apply.

We now consider whether the same natural kind terms should occur obliquely in attributions of propositional attitudes to Adam and Adam$_{te}$. Let us assume, what seems obvious, that Adam has propositional attitudes correctly attributed in English with his own (English) natural kind terms in oblique position. He hopes that there is some water within twenty miles; he believes that sailboat masts are often made of aluminium, that elms are deciduous trees distinct from beeches, that shrimp are smaller than mackerel. Does Adam$_{te}$ have these same attitudes, or at least attitudes with these same contents? As the case has been described, I think it intuitively obvious that he does not.

At least two broad types of consideration back the intuition. One is that it is hard to see how Adam$_{te}$ could have acquired thoughts involving the concept of water (aluminium, elm, mackerel).[14] There is no water on Twin-Earth, so he has never had any contact with water. Nor has he had contact with anyone else who has had contact with water. Further, no one on Twin-Earth so much as uses a word which means *water*. It is not just that water does not fall in the extension of any of the Twin-

Earthians' terms. The point is that none of their terms even translates into our (non-indexical) word 'water'. No English$_{te}$-to-English dictionary would give 'water' as the entry for the Twin-Earthians' word. It would thus be a mystery how a Twin-Earthian could share any of Adam's attitudes that involve the notion of water. They have not had any of the normal means of acquiring the concept. The correct view is that they have acquired, by entirely normal means, a concept expressed in their language that bears some striking, superficial similarities to ours. But it is different. Many people in each community could articulate things about the respective concepts that would make the difference obvious.

There is a second consideration—one that concerns truth—that backs the intuition that Adam$_{te}$ lacks attitudes involving the notion of water (aluminium, elm, mackerel). There is no water on Twin-Earth. If Adam$_{te}$ expresses attitudes that involve the concept of water (as opposed to twater), a large number of his ordinary beliefs will be false—that *that* is water, that there is water within twenty miles, that chemists in his country know the structure of water, and so forth. But there seems no reason to count his beliefs false and Adam's beliefs true (or vice versa). Their beliefs were acquired and relate to their environments in exactly parallel and equally successful ways.

The differences between the attitudes of Adam and Adam$_{te}$ derive not from differences in truth-value, but from differences in their respective environments and social contexts. They give different sorts of entities as paradigm cases of instances of the term. Their uses of the term are embedded in different communal usages and scientific traditions that give the term different constant, conventional meanings, In normal contexts, they can explicate and use the term in ways that are informative and socially acceptable within their respective communities, In doing, so, they express different notions and different thoughts with these words. Their thoughts and statements have different truth-conditions and are true of different sorts of entities.

Of course, Adam$_{te}$ believes of XYZ everything Adam believes of water—*if* we delete all belief-attributions that involve 'water' and 'twater' in oblique position, and assume that there are no relevant differences between uses of others among their natural kind terms. In a sense, they would explicate the terms in the same way. But this would show that they have the same concept only on the assumption that each must have *verbal means besides 'water'* of expressing his concept—means that suffice in every outlandish context to distinguish that concept from others. I see no reason to accept this assumption. Nor does it seem antecedently plausible.

So far I have argued that Adam and Adam$_{te}$ differ as regards the *contents* of their attitudes. This suffices to show that their mental states, ordinarily so-called, as well as the extensions of their terms differ. But the examples we used involved relational propositional attitudes: belief that *that* is water (twater), that some water (twater) is within twenty miles of *this place*, that chemists in *this country* know the structure of water (twater), and so on. Although these do not involve 'water' as an indexical expression and some are not even of (*de*) water (twater), they are, plausibly, propositional attitudes in the wide sense. Thus these examples do not strictly show that Adam and Adam$_{te}$ differ in their *de dicto* attitudes—attitudes in the narrow sense.

But other examples do. Adam might believe that some glasses somewhere sometime contain some water, or that some animals are smaller than all mackerel. Adam$_{te}$ lacks these beliefs. Yet these ascriptions may be interpreted so as not to admit of ordinary existential generalization on positions in the 'that'-clauses, and not to be *de re* in any sense. We can even imagine differences in their *de dicto* beliefs that correspond to differences in truth-value. Adam may believe what he is falsely told when someone mischievously says, 'Water lacks oxygen'. When Adam$_{te}$ hears the same words and believes what *he* is told, he acquires (let us suppose) a true belief: twater does lack oxygen.[15]

I shall henceforth take it that Adam and Adam$_{te}$ have relevant propositional attitudes from whose content ascriptions no application of existential generalization is admissible. None of these contents need be applied by the subjects—*de re*—to objects in the external world. That is, the relevant attitudes are purely *de dicto* attitudes. Yet the attitude contents of Adam and Adam$_{te}$ differ.

IV

Thus it would seem that on the construal of 'narrow sense' we have been exploring, the Twin-Earth examples fail to show that psychological states in the narrow sense (or the contents of such states) do not 'fix' the extensions of terms that Adam and Adam$_{te}$ use. For different contents and different propositional attitudes correspond to the different extensions. This conclusion rests, however, on a fairly narrow interpretation of 'fix' and—what is equally important—on a plausible, but restrictive application of 'narrow sense'. Let me explain these points in turn.

Propositional attitudes involving non-indexical notions like that of water do 'fix' an extension for the term that expresses the notion. But they do so in a purely semantical sense: (necessarily) the notion of water is true of all and only water. There is, however, a deeper and vaguer sense in which non-relational propositional attitudes do not fix the extensions of terms people use. This point concerns explication rather than purely formal relationships. The Twin-Earth examples (like the examples from 'Individualism and the Mental') indicate that the order of explication does not run in a straight line from propositional attitudes in the 'narrow sense' (even as we have been applying this expression) to the extensions of terms.[16] Rather, to know and explicate what a person believes *de dicto*, one must typically know something about what he believes *de re*, about what his fellows believe *de re* (and *de dicto*), about what entities they ostend, about what he and his fellows' words mean, and about what entities fall in the extensions of their terms.

A corollary of this point is that one cannot explicate what propositional attitude *contents* a person has by taking into account only facts about him that are non-intentional and individualistic. There is a still flourishing tradition in the philosophy of mind to the contrary. This tradition claims to explain psychological states in terms of non-intentional, functional features of the individual—with no reference to the nature of the environment or the character of the actions, attitudes, and conventions of other individuals.[17] Although there is perhaps something to be said for taking non-intentional, individualistic research strategies as one reasonable approach to explaining the behaviour of individuals, the view is hopelessly oversimplified as a philosophical explication of ordinary mentalistic notions.

Even in so far as individualism is seen as a research strategy, like the 'methodological solipsism' advocated by Jerry Fodor, it is subject to limitations. Such strategies, contrary to Fodor's presumptions, cannot be seen as providing a means of individuating ordinary ('non-transparent') attributions of content. Indeed, it is highly doubtful that a psychological theory can treat psychological states as representational at all, and at the same time individuate them in a strictly individualistic, formal or 'syntactic' way. One could, I suppose, have a theory of behaviour that individuated internal states 'syntactically'. Or one could have a representational theory (like most of the cognitive theories we have) which abstracts, in particular attributions to individuals, from the question of whether or not the attributed contents are true. But the latter type of theory, in every version that has had genuine use in psychological theory, relies on individuation of the contents, individuation which involves complex reference to entities other than the individual. Putnam's examples, interpreted in the way I have urged, constitute one striking illustration of this fact.

These remarks invite a reconsideration of the expression 'psychological state in the narrow sense'. Putnam originally characterized such states as those that do not 'presuppose' (entail) the existence of entities other than the subject. And we have been taking the lack of presupposition to be co-extensive with a failure of existential generalization and an absence of *de re* attitudes. Thus we have been taking psychological states in the 'narrow sense' to be those whose standard 'that'-clause specification does not admit of existential generalization on any of its expressions, and is not in any sense *de re*. But our weakened construal of 'fix' suggests a correction in the application of the notion of presupposition. One might say that Adam's *de dicto* attitudes involving the notion of water *do* presuppose the existence of other entities. The conditions for individuating them make essential reference to the nature of entities in their environment or to the actions and attitudes of others in the community. Even purely *de dicto* propositional attitudes presuppose the existence of entities other than the subject in this sense of presupposition. On this construal, *none* of the relevant attitudes, *de re* or *de dicto*, are psychological states in the narrow sense.

I want to spend the remainder of the section exploring this broadened application of the notion of presupposition. The question is what sorts of relations hold between an individual's mental states and other entities in his environment by virtue of the fact that the conditions for individuating his attitude contents—and thus his mental states and events—make reference to the nature of entities in his environment, or at least to what his fellows consider to be the nature of those entities.

We want to say that it is logically possible for an individual to have beliefs involving the concept of water (aluminium, elm, mackerel) even though there is no water (and so on) of which the individual holds these beliefs. This case seems relatively unproblematic. The individual believer might simply not be in an appropriately direct epistemic relation to any of the relevant entities. This is why existential generalization can fail and the relevant attitudes can be purely *de dicto*, even though our method of individuating attitude contents makes reference to the entities.

I think we also want to say something stronger: it is logically possible for an individual to have beliefs involving the concept of water (aluminium, and so on), even though there exists no water. An individual or community might (logically speaking) have been wrong in thinking that there was such a thing as water. It is epistemically possible—it might have turned out— that contrary to an individual's beliefs, water did not exist.

Part of what we do when we conceive of such cases is to rely on actual circumstances in which these illusions do not hold— rely on the actual existence of water—in order to individuate the notions we cite in specifying the propositional attitudes. We utilize—must utilize, I think—the actual existence of physical stuffs and things, or of other speakers or thinkers, in making sense of counterfactual projections in which we think at least some of these surroundings away.

But these projections are not unproblematic. One must be very careful in carrying them out. For the sake of argument, let us try to conceive of a set of circumstances in which Adam holds beliefs he actually holds involving the notion of water (aluminium etc.), but in which there is no water and no community of other speakers to which Adam belongs. Adam may be deluded about these matters: he may live in a solipsistic world. What is problematic about these alleged circumstances is that they raise the question of how Adam *could* have propositional attitudes involving the notion of water. How are they distinguished from attitudes involving the notion of twater, or any of an indefinitely large number of other notions?

In pressing this question, we return to considerations regarding concept acquisition and truth. How, under the imagined circumstances, did Adam acquire the concept of water? There is no water in his environment, and he has contact with no one who has contact with water. There seems no reason derivable from the imagined circumstances (as opposed to arbitrary stipulation) to suppose that Adam's words bear English interpretations instead of English$_{te}$ interpretations, since there are no other speakers in his environment. Nothing in Adam's own repertoire serves to make 'water' mean *water* instead of *twater*, or numerous other possibilities. So there seems no ground for saying that Adam *has* acquired the concept of water.

Considerations from truth-conditions point in the same direction. When Adam's beliefs (as held in the putative solipsistic world) are carried over to and evaluated in a 'possible world' in which twater (and not water) exists, why should some of the relevant beliefs be false in this world and true in a world in which water exists, or vice versa? Nothing in the solipsistic world seems to ground any such distinction. For these reasons, it seems to me that one cannot credibly imagine that Adam, with his physical and dispositional life history, could have beliefs involving the notion of water, even though there were no other entities (besides his attitude contents) in the world.

We have now supported the view that the point about explication and individuation brings with it, in this case, a point about entailment. Adam's psychological states in the narrow sense (those that do not entail the existence of other entities) do not fix (in either sense of 'fix') the extensions of his terms. This is so not because Adam's beliefs involving the notion of water are indexical or *de re*, and not because he has the same propositional attitudes as Adam$_{te}$ while the extensions of his terms differ. Rather it is because all of Adam's attitude contents involving relevant natural kind notions—and thus all his relevant attitudes (whether *de re* or *de dicto*)—are individuated, by reference to other entities. His having these attitudes in the relevant circumstances entails (and thus presupposes in Putnam's sense) the existence of other entities.

The exact nature of the relevant entailment deserves more discussion than I can give it here. As I previously indicated, I think that Adam's having attitudes whose contents involve the notion of water does not entail the existence of water. If by some wild communal illusion, no one had ever really seen a relevant liquid in the lakes and rivers, nor had drunk such a liquid, there might still be enough in the community's talk to distinguish the notion of water from that of twater and from other candidate notions. We would still have our chemical analyses, despite the illusoriness of their object. (I assume here that not *all* of the community's beliefs involve similar illusions.)[18] I think that Adam's having the relevant attitudes probably does not entail the existence of other speakers. Prima facie, at least, it would seem that if he did interact with water and held a few elementary true beliefs about it, we would have enough to explain how he acquired the notion of water—enough to distinguish his having that notion from his having the notion of twater. What seems incredible is to suppose that Adam, in his relative ignorance and indifference about the nature of water, holds beliefs whose contents involve the notion, even though neither water nor communal cohorts exist.

V

It should be clear that this general line promises to have a bearing on some of the most radical traditional sceptical positions. (I think that the bearing of the argument in 'Individualism and the Mental' is complementary and more comprehensive.) The line provides fuel for the Kantian strategy of showing that at least some formulations of traditional scepticism accept certain elements of our ordinary viewpoint while rejecting others that are not really separable. Exploring the epistemic side of these issues, however, has not been our present task.

Our main concern has been the bearing of these ideas on the philosophy of mind. What attitudes a person has, what mental events and states occur in him, depends on the character of his physical and social environment. In some instances, an individual's having certain *de dicto* attitudes *entails* the existence of entities other than himself and his attitude contents. The Twin-Earth thought-experiment may work only for certain propositional attitudes. Certainly its clearest applications are to those whose contents involve non-theoretical natural kind notions. But the arguments of 'Individualism and the Mental' suggest that virtually no propositional attitudes can be explicated in individualistic terms. Since the intentional notions in terms of which propositional attitudes are described are irreducibly non-individualistic, no purely individualistic account of these notions can possibly be adequate.

Although most formulations of the lessons to be learned from Twin-Earth thought-experiments have seemed to me to be vague or misleading in various ways, many of them indicate a broad appreciation of the general drift of the argument just presented. In fact, as I indicated earlier, the general drift is just beneath the surface of Putnam's paper. A common reaction, however, is that if our ordinary concept of mind is non-individualistic, so much the worse for our ordinary concept.

This reaction is largely based on the view that if mentalistic notions do not explain 'behaviour' individualistically, in something like the way chemical or perhaps physiological notions do, they are not respectable, at least for 'cognitive' or 'theoretical' as opposed to 'practical' purposes. I cannot discuss this view in the detail it deserves here. But it has serious weaknesses. It presupposes that the only cognitively valuable point of applying mentalistic notions is to explain individual 'behaviour'. It assumes that the primary similarities in 'behaviour' that mentalistic explanations should capture are illustrated by Adam's and Adam$_{te}$'s similarity of physical movement. It assumes that there are no 'respectable' non-individualistic theories. (I think evolutionary biology is a counterexample—not to appeal to much of cognitive psychology and the social sciences.) And it assumes an unexplicated and highly problematic distinction between theoretical and practical purposes. *All* of these assumptions are questionable and, in my view, probably mistaken.

The non-individualistic character of our mentalistic notions suggests that they are fitted to purposes other than (or in addition to) individualistic explanation. The arguments I have presented, here and earlier, challenge us to achieve a deeper understanding of the complex system of propositional attitude attribution. The purposes of this system include describing, explaining, and assessing people and their historically and socially characterized activity against a background of objective norms—norms of truth, rationality, right. Some form of fruitful explanation that might reasonably be called 'psychological' could, conceivably, ignore such purposes in the interests of individualistic explanation. But animus against mentalistic notions because they do not meet a borrowed ideal seems to me misplaced. That, however, is a point for sharpening on other occasions.

Notes to chapter 2

¹ It is difficult to avoid at least a limited amount of philosophical jargon in discussing this subject. Since much of this jargon is subjected to a variety of uses and abuses, I will try to give brief explications of the most important special terms as they arise. An ordinary-language discourse is *intentional* if it contains *oblique* occurrences of expressions. (Traditionally, the term 'intentional' is limited to mentalistic discourse containing obliquely occurring expressions, but we can ignore this fine point.) An *oblique* (sometimes 'indirect' and less appropriately, 'opaque' or 'non-transparent') occurrence of an expression is one on which *either* substitution of co-extensive expressions may affect the truth-value of the whole containing sentence, *or* existential generalization is not a straightforwardly valid transformation. For example, 'Al believes that many masts are made of aluminium' is intentional discourse (as we are reading the sentence) because 'aluminium' occurs obliquely. If one substituted 'the thirteenth element in the periodic table' for 'aluminium', one might alter the truth-value of the containing sentence.

The characterization of *de re attitudes* (sometimes 'relational attitudes') is at bottom a complex and controversial matter. For a detailed discussion, see my 'Belief De Re' (1977, pp. 338–62), especially section I. For present purposes, we shall say that *de re* attitudes are those where the subject or person is unavoidably characterized as being in a not-purely-conceptual, contextual relation to an object (*re*), *of* which he holds his attitude. Typically, though not always, a term or quantified pronoun in non-oblique position will denote the object, and the person having the attitude will be said to believe (think, etc.) that object to be φ (where 'φ' stands for oblique occurrences of predicative expressions). *De re* attitudes may equivalently, and equally well, be characterized as those whose content involves an ineliminable indexical element which is applied to some entity. *De dicto* attitudes (sometimes 'notional attitudes') are those that are not *de re*.

[2] An attitude *content* is the semantical value associated with oblique occurrences of expressions in attributions of propositional attitudes. Actually, there may be more to the content than what is attributed, but I shall be ignoring this point in order not to complicate the discussion unduly. Thus the content is, roughly speaking, the conceptual aspect of what a person believes or thinks. If we exclude the *res* in *de re* attitudes, we may say that the content is what a person believes (thinks, etc.) in *de re* or *de dicto* attitudes. We remain neutral here about what, ontologically speaking, contents are.

[3] Hilary Putnam, 'The Meaning of "Meaning"' (reprinted in Putnam 1975b, to which page numbers refer).

[4] This point is entirely analogous to the familiar point that *knowing* is not a mental state. For knowledge depends not only on one's mental state, but on whether its content is true. The point above about indexicals and mental states is the analogue for predication of this traditional point about the relation between the mind and (complete) propositions. Neither the truth or falsity of a content nor the 'truth-of-ness' or 'false-of-ness' of a content, nor the entities a content is true of, enters directly into the individuation of a mental state. For more discussion of these points, see Burge (1979a) section IId.

[5] This rejection is logically independent of rejecting the view that the intentional can be accounted for in terms of the non-intentional. I reject this view also. But here is not the place to discuss it.

[6] Burge (1979a). Much of the present paper constitutes an elaboration of remarks in footnote 2 of that paper.

[7] Anyone who wishes to resist our conclusions merely by claiming that XYZ is water will have to make parallel claims for aluminium, helium, and so forth. Such claims, I think, would be completely implausible.

[8] I am omitting a significant extension of Putnam's ideas. Putnam considers 'rolling back the clock' to a time when everyone in each community would be as ignorant of the structure of water and twater as Adam and Adam$_{te}$ are now. I omit this element from the thought-experiment, partly because arriving at a reasonable interpretation of this case is more complicated and partly because it is not necessary for my primary purposes. Thus, as far as we are concerned, one is free to see differences in Adam and Adam$_{te}$'s mental states as deriving necessarily from differences in the actions and attitudes of other members in their respective communities.

[9] This view and the one described in the following sentence were anticipated and criticized in Burge (1979a), section IId and note 2. Both views have been adopted by Jerry Fodor (1980, pp. 63–73). I believe that these views have been informally held by various others. Colin McGinn (1977, pp. 521–35), criticizes Putnam (correctly, I think) for not extending his theses about meaning to propositional attitudes. But McGinn's argument is limited to claiming that the *res* in relational propositional attitudes differ between Earth and Twin-Earth and that these *res* enter into individuating mental states. (The congeniality of this view with Putnam's claim that natural kind terms are indexical is explicitly noted.) Thus the argument supports only the position articulated in the first paragraph of this paper and is subject to the deflationary interpretation that followed (cf. also note 4). McGinn's argument neither explicitly accepts nor explicitly rejects the position subsequently adopted by Fodor and cited above.

[10] Putnam (1974), p. 451. Cf. also p. 277 of 'Language and Reality' (Putnam 1975b).

[11] In explaining the traditional assumption of 'methodological solipsism', Putnam writes: 'This assumption is the assumption that no psychological state, properly so-called, presupposes the existence of any individual other than the subject to whom that state is ascribed. (In fact, the assumption was that no psychological state presupposes the

existence of the subject's *body* even: if P is a psychological state, properly so-called, then it must be *logically possible* for a "disembodied mind" to be in P.)' (1975b, p. 220, the second italics mine). He also gives examples of psychological states in the 'wide sense', and characterizes these as *entailing* the existence of other entities besides the subject of the state (ibid., p. 220). Although there is little reason to construe Putnam as identifying entailment and presupposition, these two passages taken together suggest that for his purposes, no difference between them is of great importance. I shall proceed on this assumption.

    12 This guess is Fodor's (1980). As far as I can see, the interpretation is not excluded by anything Putnam says. It is encouraged by some of what he says—especially his remarks regarding indexicality and his theory about the normal form for specifying the meaning of 'water'.

    13 Jerry Fodor (1980) states that inhabitants of Twin-Earth harbour the thought that water is wet—even granted the assumption that there is no water on Twin-Earth and the assumption that their thought is not 'about' water (H$_2$O), but about XYZ. Fodor provides no defence at all for this implausible view.

    14 More jargon. I shall use the terms '*concept*' and '*notion*' interchangeably to signify the semantical values of obliquely occurring parts of content-clauses, parts that are not themselves sentential. Thus a concept is a non-propositional part of a content. The expressions 'concept' and 'notion' are, like 'content', intended to be ontologically neutral. Intuitively, a concept is a context-free way a person thinks about a stuff, a thing, or a group of things.

    15 As I mentioned earlier, one might hold that 'water' names an abstract property or kind and that attitude attributions typically attribute *de re* attitudes *of* the kind. I do not accept this view, or see any strong reasons for it. But let us examine its consequences briefly. I shall assume that 'kind' is used in such a way that water is the same kind as H$_2$O. To be minimally plausible, the view must distinguish between kind and concept (cf. note 14)—between the kind that is thought of and the person's way of thinking of it. 'Water' may express or indicate one way of thinking of the kind—'H$_2$O' another. Given this distinction, previous considerations will show that Adam and Adam$_{te}$ apply different concepts (and contents) to the different kinds. So even though their attitudes are not 'narrow', they still have different mental states and events. For mental states and events are individuated partly in terms of contents.

    16 I am tempted to characterize this as Putnam's own primary point. What counts against yielding to the temptation is his interpretation of natural kind terms as indexical, his focus on meaning, and his statement that those on Twin-Earth have the same thoughts (1975b, p. 224) as those on Earth. Still, what follows is strongly suggested by much that he says.

    17 Works more or less explicitly in this tradition are Putnam, 'The Nature of Mental States' in Putnam (1975b), p. 437; Harman (1973), pp. 43–6, 56–65; Lewis (1972b), pp. 249–50 (1974), pp. 331ff; Fodor (1975), chapter I, and 1980.

    18 Thus I am inclined to think that, if one is sufficiently precise, one could introduce a 'natural kind' notion, like water without having had any causal contact with instances of it. This seems to happen when chemical or other kinds are anticipated in science before their discovery 'in nature'. The point places a *prima facie* limitation on antisceptical uses of our argument. Thus . . . I have been careful to emphasize Adam's relative ignorance in our criticism of solipsistic thought experiments.

3

De re Belief and Methodological Solipsism

Kent Bach

If we had no *de re* beliefs about things in the world, our knowledge of them would be by description only. Thinking of an object would never involve actually having that object in mind but would consist merely in using some individual concept satisfied by that object alone.[1] Had some other object been the sole satisfier of that concept, one would have been thinking about that object instead. Having a *de re* belief about an object requires being in one of a certain class of relations to that object, and it is the job of a theory of *de re* belief to characterize that class. No general theory of *de re* relations has yet been proposed, and I will not offer one here. However, I will present a proposal for the special and fundamental case of perceptual belief, after mentioning some features of *de re* beliefs in general. Recognizing these features will dispel certain misconceptions that have made *de re* beliefs seem inherently problematic or even downright impossible.

You currently believe, at least now that I mention it, that the page you are reading is rectangular and white, with black print. Reflecting on such a belief, no one uncorrupted by philosophy would suppose that it, much less all his beliefs about physical objects, is merely descriptive. And if he read the opening pages of Strawson's *Individuals* (1959), he would find good reason for not so supposing. Yet some philosophers find *de re* beliefs puzzling, sometimes so much as to want to reduce them to descriptive beliefs[2] or even to banish them altogether. If *de re* beliefs were, like other beliefs, propositional attitudes, such a move would be well motivated. For then their contents could only be singular (Kaplan 1978) or *de re* propositions (Perry 1979), propositions literally including the objects they are about, not just containing symbols for those objects. If there were such propositions and they could be contents of psychological states, their objects would literally be parts of those mental contents. No wonder that Russell, who held that the basic constituents of propositions must be objects of acquaintance (in his special sense), could not countenance *de re* propositions about physical objects and hence had to regard our knowledge of physical objects as by description only. However, as we will see in discussing the general nature of *de re* beliefs, their contents are not propositions of any kind, contrary to what most complaints about them presuppose. As Tyler Burge (1977) puts it, a *de re* belief is not 'fully conceptualized', but 'is a belief whose correct ascription places the believer in an appropriate nonconceptual relation to objects the belief is about' (p. 346). Regrettably, subsequent critics of *de re* belief have continued to construe it as propositional and have not countered Burge's arguments that *de re* beliefs are epistemologically more fundamental than (pp. 347–50), and are not reducible to (pp. 350–3), descriptive (*de dicto*) beliefs. In seeking to define the nature of *de re* beliefs generally and to characterize the contents of perceptual beliefs in particular, I am heavily indebted to Burge's pioneering work.

Dispelling misconceptions about *de re* beliefs might render them philosophically acceptable but for their seeming incompatibility with methodological solipsism in psychology (MS). Hilary Putnam coined that phrase for 'the assumption that no psychological state, properly so-called, presupposes the existence of any individual other than the subject to whom that state is ascribed' (1975a, p. 136). Thus, knowing and remembering (or any state expressed by a factive verb) are psychological states only in 'the wide sense', unlike believing and recalling, which are psychological states in the narrow sense. For Putnam wide states are 'hybrid states', combining a narrow state with something extrapsychological.[3] Now *de re* beliefs, assuming them not to be reducible to descriptive beliefs, seem not to fall neatly into either category. For example, a perceptual belief about some physical object seems not to be narrow, at least not if the object is somehow essential to the belief, but it is not straightforwardly wide either. Suppose it were a wide, hybrid state. Then presumably it could be factored into a narrow state plus something extrapsychological, namely its object. Call the object *a* and suppose the belief to be that *a* is F. It seems that factoring *a* out of the belief leaves not merely a narrow state but an incomplete state, for we are assuming, in taking the belief not to be reducible to a descriptive belief, that it contains no individual concept satisfied solely by *a*. Surely what remains as its content is not merely the existential proposition that something is F, but then how is *a* represented in that content if not by an individual concept?[4] That is the question that will be answered by the account of perceptual belief to be proposed here, in a way that reconciles with MS the claim that perceptual beliefs are psychological states properly so-called.

I. Two Versions of Methodological Solipsism

To motivate this effort we need to understand why MS *should* constrain our conception of genuinely psychological states, but as we will soon see, there is an important unclarity about the strength of this constraint. As the archadvocate of MS Jerry Fodor (1980) emphasizes, the psychological point of MS, as opposed to its foundationalist point in epistemology, is that what is outside a person's mind is irrelevant to psychology. Regardless of how the world is in comparison to how it is represented as being and regardless of how it may change while the person's psychological states remain the same, everything is the same as far as psychology is concerned. According to MS, the psychologically appropriate way to individuate types of beliefs, desires, and other intentional states is by their contents, not by relations the subject has to things in the world, which could have been different without affecting content. When we cite such states to explain or predict actions, we specify the representational contents of those states. What matters is how the subject represents the world, not how the world actually is. Accordingly, any reasonable formulation of laws and other generalizations about mental states should individuate types of states by their contents. Besides, according to Fodor, the only alternative is 'naturalistic' psychology and it is utterly unfeasible. It would deal with relations between persons and things in the world (including other persons), but there is not the slightest reason to suppose, at least until our knowledge approaches that of Laplace's demon, that there will be discovered 'law instantiating descriptions' under which 'objects of thought enter into scientific generalizations about the relations between thoughts and their objects'. Rather, all 'we can reasonably hope for is a theory of mental states opaquely individuated'. If by 'opaquely individuated' Fodor meant 'individuated by contents', there would be nothing ambiguous in his formulation of the constraints imposed on psychology by MS. But it is not at all clear that this is what he means.

Among the psychological states at issue are those whose contents are commonly taken to be propositions[5] and which can be regarded, as Russell first called them, as propositional attitudes. They are thus relations to propositions, but if what Fodor calls the 'representational theory of the mind' is correct, these relations are realized, as a matter of empirical fact, by relations to internal representations. They are not identical to those relations, for whereas the relata of propositional attitudes are contents (propositions), the relata of the relations that realize them are representations, which are not contents but (at best) have contents. Whether internal representations do have contents and whether psychology need suppose that they do are two hotly debated questions about the representational theory of the mind. Fodor frequently attributes contents to them, but nevertheless he maintains that the only kind of psychology that is feasible is not naturalistic psychology, which requires a theory of content, but 'computational psychology', which does not. From the computational perspective, propositional attitudes are realized by computational relations to representations, and these are individuated by the forms (syntactic properties) of those representations. As Fodor aptly makes the contrast, 'being syntactic is a way of not being semantic'. However, as Stephen Schiffer has since argued (1980), a theory of psychological states concerned only with their 'internal functional roles' would not suffice as a theory of their contents, and since these roles are strictly computational, defined over the forms of representations, these representations might as well be construed by psychology as uninterpreted symbols. Their putative contents play no internal functional role.

Schiffer's position shows that there are two ways of taking MS. One way of taking it is that psychology should work on the assumption that propositional attitudes are realized by relations to symbols whose only relevant properties are formal. This austere interpretation of MS might be called *formal* methodological solipsism (FMS). The weaker version insists that internal representations should be treated as having not only formal but also conceptual[6] properties, which are not reducible to or fully explainable by formal properties. This liberal form of MS, call it *conceptual* methodological solipsism (CMS), leaves open the nature of these conceptual properties. They might be Fregean senses or perhaps the kinds of properties that Katz (1978) assigns to semantic representations. The only requirement is that they be specifiable without reference to anything external to the person whose states they characterize.

Does Fodor favour FMS over CMS? It is not clear. The unclarity is due partly to Fodor's discussion of semantic properties. When he first mentions them, he includes meaning along with truth and reference, but thereafter he neglects meaning altogether. In denying the feasibility of naturalistic psychology, he expresses scepticism that psychology could succeed in ascribing semantic properties to internal representations, and yet the only semantic properties he mentions are truth and reference. Of course these extensional properties require a naturalistic psychology for their attribution, but what about intensional properties? Fodor says nothing about the kind of psychology required for their attribution, and seems to assume that the only alternative to a naturalistic psychology is a computational one. However, and this is the other source of unclarity about his position, he does not require computational psychology to be strictly computational. In characterizing the computational theory of the mind, he says that 'content alone cannot distinguish thoughts' and that 'two thoughts can be distinct in content only if they can be identified with relations to formally distinct representations'. Clearly this 'formality condition' does not entail FMS and is perfectly compatible with CMS. It allows that two formally distinct representations could nevertheless be semantically identical, so that they could have the same internal functional role (as determined by their identical contents opaquely individuated) while being formally distinct.

I will suppose that Fodor's methodological solipsism is conceptual rather than formal, even though he says nothing about whether psychology can and should be more than computational without being naturalistic. He does not explicitly reject the possibility of a theory of intensional content, and his extensive discussion of opaque construals and opaque individuation of psychological states would make no sense if internal representations need be regarded merely as uninterpreted symbols. After all, the transparent/opaque distinction can be drawn only in terms of reference and truth-value. Indeed, to call the austere, strictly formal version of MS 'methodological solipsism' is, at best, ironical.

In any case it seems that FMS is too strong, notwithstanding Schiffer's powerful arguments in its favour, for surely formally distinct representations can be semantically identical, e.g., 'Water is thinner than blood' and 'Blood is thicker than water'. For our purposes the operative version of methodological solipsism will be CMS, not that there are no outstanding problems with it. For our examination of *de re* beliefs, whose contents are not fully conceptual, the problem of how to individuate their contents opaquely, thereby satisfying the demands of MS, would not even arise if, as FMS requires, psychological states are to be treated strictly computationally, hence not to be construed as having contents at all. Our problem is to formulate a conception of *de re* beliefs on which they can be construed as psychological states properly so-called, and yet do justice to the intuitive conception of them as essentially related to their objects.

Accepting CMS as a working assumption, I am supposing that a psychological theory of content is possible without being naturalistic. I agree with Fodor and Schiffer that a theory of extensional content, i.e., a theory of truth and reference for internal representations, is practically out of the question and not properly psychology anyway. However, there remains the possibility of a conceptual psychology, one that appeals to senses or intensions, coupled with the kind of computational psychology envisioned by Fodor and Schiffer. As Katz (1978) has argued in regard to the theory of meaning for natural languages, so we may argue that the theory of content for internal representations requires the notions of synonymy, redundancy, informal entailment, and possibly ambiguity.[7] If we accept not only Katz's demand but Schiffer's assumption that the theory of content for public language reduces completely to the theory of content for thought, as I do (Bach and Harnish 1979), these requirements follow immediately. However, Burge (1979a) has produced a series of arguments that can be taken as showing that conceptual psychology is not compatible with MS.

If Burge is correct, then either CMS must yield to FMS or methodological solipsism must be abandoned altogether. Against what might be called psychological individualism, against the possibility of a psychologically autonomous theory of conceptual contents, Burge adapts Putnam's (1975a) arguments regarding general terms to conceptual contents. Although I do not find Burge's arguments ultimately convincing—I confess to a deep-seated, Cartesian conviction that my psychological states could be just as they are, having just the contents they have, regardless of how the world is—they are well worth serious examination.[8] I mention Burge's anti-individualism here not to criticize it but to note a rather subtle difference between the challenge it poses to CMS and the one we will consider posed by *de re* beliefs. Burge's challenge relies only on cases of general, purely conceptual belief and seeks to overturn the presupposition of CMS that conceptual contents are a matter of internal psychological state, whose attribution is independent of how the external world, physical and social, actually is. Burge uses his examples to show that two people could have different beliefs while being in the same psychological state *narrowly construed*, and vice versa. He argues that ascriptions of content, though genuinely psychological, have an ineliminable social dimension, but rather than opt for FMS, whose conception of the psychological he regards as unduly restrictive, he rejects psychological individualism and, by implication, MS in either form. In other words, two people who have different beliefs are in different psychological states properly so-called, even if they are in the same psychological states narrowly construed. Moreover, they could be in different psychological states in the narrow sense and yet have the same beliefs, thereby being in the same psychological state properly construed.[9]

Now Burge does not deny the possibility of a narrow psychology, only its interest (1979a, pp. 105–9), so we might suppose that he would allow that types of psychological states can be narrowly individuated, though not in terms of their contents since content is not a narrow-psychological property. Then the upshot of his view is that the narrow-psychological properties of a belief do not in general determine its truth-condition. Although the sort of belief under consideration is purely conceptual, so that its content determines its truth-condition, its narrow-psychological properties neither include nor uniquely determine its content. And as Burge has since pointed out (this volume), it is not part of his position, in contrast to Putnam's, that some concepts (notably of natural kinds) are implicitly indexical.

Thus, Burge's challenge to CMS is not that content does not determine truth-condition but that content is not a narrow-psychological property. *De re* beliefs pose a different challenge. Even if they are not construed as essentially involving their objects, in which case they could not be regarded by MS as psychological states properly so-called, still they are not purely conceptual but, as I will argue, inherently indexical. Although their objects are not part of their contents, how they are related to their objects is part of their contents. So if we assume, contrary to Burge's view about conceptual contents, that content is a narrow-psychological property, *de re* beliefs have truth-conditions which are not determined solely by their narrow-psychological properties, not because their contents are not narrow but because, since they are indexical, their contents do not determine their truth-conditions. Thus, Burge's challenge in regard to conceptual contents and the challenge posed to CMS by *de re* beliefs are distinct and, I believe, independent. So in assuming rather than arguing that conceptual contents are narrowly psychological, I mean merely to facilitate discussion of the nonconceptual element that Burge himself (1977) was first to recognize as distinctive of *de re* beliefs.

II. A Conception of *De Re* Belief

The main task of a general theory of *de re* belief (and of other *de re* attitudes and episodes) is to characterize the relations that can obtain between a person and an object such that he can believe something *de* that object. Apart from the basic case of perceptual belief taken up in section III, I will not try to characterize the class of *de re* relations but will offer a general conception of *de re* beliefs. This will require explaining how they can be essentially relational without being essentially of their objects. The best way to motivate this conception is first to mention four common misconceptions about them, which have made them seem if not downright impossible at least problematic, especially given the constraints of methodological solipsism.

What De Re *Beliefs Are Not*

(1) Perhaps the most widespread misconception is that *de re* beliefs are what *de re* ascriptions ascribe, while *de dicto* (descriptive) beliefs are what *de dicto* ascriptions ascribe. This misconception can take two forms, that (a) 'believes-that' locutions can be used literally only to ascribe *de dicto* beliefs, while 'believes-of' (or 'believes-to-be') locutions can be used literally only to ascribe *de re* beliefs, and that (b) whether a singular term in the 'that'-clause of an ascription occurs opaquely or transparently determines whether the belief ascribed is *de dicto* or *de re*. But consider sentences like the following.

(T) Jill believes that the boy/Jack/he fell down.
(O) Jill believes of the boy/Jack/him that he fell down.

(T) can be used to ascribe a *de re* belief about Jack even when the description 'the boy' is used to refer to Jack. (O) can be used, even with 'Jack' or 'him', to ascribe a *de dicto* belief, for the ascriber can refer to Jack without specifying the individual concept under which Jill thinks of Jack. The singular term appearing in the 'that'-clause of a belief-ascription typically expresses part of the content of the belief only if it occurs opaquely (names and indexicals generally occur transparently, but not always), but even if it does occur opaquely, the belief ascribed need not be *de dicto*. Its occurring opaquely means only that the ascriber is not using it to refer, and not that it expresses part of the content of the belief. And even if the singular term occurs transparently, the belief ascribed could still be *de dicto*. Using the singular term transparently means only that the ascriber is using it to refer. Indeed, as Loar (1972) has observed, the ascriber can use a definite description as a singular term in the 'that'-clause both to refer to an object and to express how the believer thinks of the object.[10] As for 'believes-of' ascriptions using sentences like (O), it is true that the singular term normally occurs transparently and is used by the ascriber to refer, in this case to Jack, but that in no way implies that the belief ascribed is *de re*. It implies only that the ascriber has not attempted to give the full content of the belief, by failing to indicate how Jill thinks of Jack, be it with some individual concept or in some *de re* way.

The moral of this very short story is that different forms of belief-sentences do not mark differences in kinds of belief. Neither the grammatical form of such a sentence nor how a singular term occurs in it determines the kind of belief ascribed; the indeterminacies in how singular terms occur and in kind of belief ascribed, as Kempson (1979) has argued, are not matters of *semantic* ambiguity.[11] What the ascriber means is underdetermined by the meaning of the sentence he uses and must be inferred by his audience, perhaps in the pattern delineated by Bach and Harnish (1979). In general, using a particular form of belief-sentence does not make explicit whether the ascribed belief is *de dicto* or *de re* or whether any of the singular terms occurring in it express part of the content of the belief. There is one obvious exception: one can be explicit about the kind of ascribed belief and its content by *saying* that it is *de dicto* or *de re*, as the case may be, and that its content is such and such.[12]

Differences in belief-sentences do not shed light on differences in beliefs. Accordingly, we should be wary of efforts to explain one in terms of the other, as exemplified by numerous discussions of 'quantifying in' (to belief-contexts). However, we should not be drawn to the conclusion that the distinction between *de re* and *de dicto* beliefs is bogus, arising, as Searle thinks (1979b, p. 157), 'from a confusion between features of reports of beliefs and features of the beliefs being reported'. Also, we should be careful about how we take Fodor's (1980) glosses on MS, e.g., that 'what we can reasonably hope for is a theory of mental states opaquely individuated' and that there is a 'correspondence between narrowness and opacity on the one hand and width and transparency on the other'. Otherwise, there will be something wrong with *our* glosses. His emphasis on 'opaque attributions' and 'opaque construals' is about belief-reports, not beliefs reported, and thus misses the point, his point, that CMS requires beliefs to be type-individuated by their contents (assuming that contents are narrow-psychological properties).

(2) It is a mistake to think of *de re* beliefs as propositional attitudes, with propositions of some special kind as their contents. As Schiffer (1978) has argued, the complete content of a *de re* belief cannot be a singular proposition (Kaplan 1978), or what Perry (1979) has since called a *de re* proposition, literally containing the object the belief is about, for there must be some mode of presentation of the object. Besides, if *de re* beliefs are to be reconciled with MS, their contents cannot include their objects at all (if external to the believer). In his compelling vindication of *de re* belief, Burge (1977) denies that they are propositional, because they are not 'fully conceptualized'. They do not contain singular terms or 'thought symbols' that denote, in the strict context-independent sense, the objects they are about. However, if such symbols pick out objects 'only relative to a context, the content of the believer's attitude does not depend purely on what is expressed (*dictum*) by his symbols, or on the nature of his concepts' (p. 351). Schiffer (1977) makes the same point about the occurrence of names in belief-reports and their mental counterparts in belief-contents. The name of an object does not determine its mode of presentation in a belief about it and different beliefs about it, even under the same name, can be under different modes of presentation. Kripke's 'puzzle about belief' (1979) is a mirage that appears only when mode of presentation is neglected. Dennett's amusing Ballad of Shakey's Pizza Parlor (this volume) seems to make trouble for *de re* beliefs

only because, aside from being construed as propositional, Dennett's examples lack modes of presentation.[13]

(3) Sometimes it is supposed that having a *de re* belief about an object entails being in some special cognitive relation to it. Perhaps inspired by Russell's distinction between knowledge by acquaintance and knowledge by description, philosophers have described what they take to be this relation in various picturesque ways. Kaplan (1968) calls it 'rapport', Kim (1977) 'direct cognitive contact', and Chisholm (1980) 'epistemic intimacy'. Both Chisholm and Pollock (1980) have proposed detailed accounts of what they believe are necessary and sufficient epistemic conditions on thinking about an object in what Pollock calls a '*de re* way'. I cannot give them the detailed examination they deserve, but at least I can explain why I take epistemic approaches to *de re* belief to be mistaken in principle or, at any rate, incompatible with the relational conception endorsed by Burge and to be elaborated here. As our discussion of perceptual belief will make evident, it is possible to have a *de re* belief about an object one is looking at, that it is red, say, and, after that object has been instantaneously and surreptitiously replaced by another, to have the same *de re* belief about the second object. The belief-token remains the same psychologically—its content is unchanged—but, since its truth-condition is now that the second object rather than the first be red, it is semantically different. Because of this possibility, I reject any epistemic condition on *de re* belief, at least as I construe it. Similarly, having a *de re* belief, about an object does not require knowing what it is or being able to identify it. Such a requirement is not only vague but context-dependent (Boer and Lycan 1975). Moreover, including it in an account of *de re* belief would make the mistake, mentioned under (2) above, of supposing that the objects of *de re* beliefs are determined by elements of their contents. However, the content of a *de re* belief is essentially indexical—tokens of the same type (with the same content) can have different objects—and its object is determined contextually. As McGinn so aptly puts it (this volume), the object 'is determined by the occurrence of a representation in a context, not by way of a representation *of* the context'. As we will see, the contextual relation that determines

the object of a *de re* belief (token) is not epistemic. Besides, the epistemic relations discussed by the writers mentioned above are proposed not as what determines the object of a *de re* belief but as what determines that a belief about an object is *de re*.

(4) That *de re* beliefs are relational has led people to think that a *de re* belief is essentially about an object in a way that a descriptive belief is not. As with many essentialist claims, this is true in one respect but false in another. The two should not be confused, lest we fall back into one of the misconceptions that make *de re* beliefs seem irreconcilable with MS or otherwise problematic. Contrast the descriptive belief that the F is G, where 'the F' is uniquely satisfied, with the *de re* belief of the F (under some mode of presentation) that it is G. They are both about the same object and they have the same predicative content ('is G').[14] Clearly the descriptive belief is not essentially about its object, since some object other than the actual F might have been the only F instead, in which case the belief would have been about that object. Its object is whatever happens to satisfy uniquely the element of its content expressed by 'the F'. However, the object of the *de re* belief (token) is determined not by any element in its content but by the context of belief. The relevant part of the content, the mode of presentation, determines what that contextual relation is, not the relatum of that relation. If belief-type is individuated by content (mode of presentation together with predicative content), as it should be for psychological purposes, different tokens of the same type (as noted under (3) above, even the same token at different times) could be related to different objects. Thus, as type-individuated psychologically, *de re* beliefs are no more essentially about their objects than are descriptive beliefs.

However, there is a way in which they are essentially about their objects. From a semantic viewpoint they can be individuated by their truth-conditions, whose determination is context-relative. In the context of a specific *de re* belief-token, wherein its object is determined, its truth-condition is that that very object be G. Its content does not determine that object, for in that case the object would be whatever uniquely satisfies that content element, and so there is no question of whether its content could have determined some other object instead. Semantically, the appropriate way to type-individuate *de re* beliefs is by object and predicative content, and relative to this way of type-individuating them they are essentially about their objects. However, they can be psychologically distinct if their objects are presented in different ways, just as two psychologically identical *de re* beliefs can be semantically distinct if they have different objects.

Distinguishing psychologically from semantically motivated ways of individuating types of beliefs provides a ready reply to Stich's (1978b) and Perry's (1979) denial that belief is a properly psychological notion. They rely on the arbitrary assumption that beliefs are or ought to be type-individuated only by truth-conditions, but surely beliefs, like anything else, can be type-individuated in different ways for different purposes. Rather than conclude that for the purposes of explanation and prediction psychology does not need belief, they should have concluded that psychology should individuate types of beliefs in a way suitable to those purposes, viz., by contents. People's behaviour is intelligible only if we understand how they represent the world and particular things in it, and that is a matter of ascertaining the contents of their beliefs and other attitudes.

What de re *beliefs are*

There are three complementary ways of characterizing the difference between a *de re* belief of the F (presented in a certain way) that it is G and the corresponding descriptive belief that the F is G. The *de re* belief is (A) *relational*, (B) *indexical*, and (C) *incompletely representational*. More precisely, the way in which its object is determined is relational (rather than satisfactional), its form is indexical, and its content is incompletely representational. Consider each feature in turn.

(A) The way in which the object of a *de re* belief-token is determined is relational. Its object is that which stands to it in a certain relation, as determined by the mode of presentation in its content. Thus, its content determines that relation but not the object which stands in that relation to it. Different objects can stand in the same relation to different tokens of the same belief-type. In contrast, the object of a belief-token that the F is G is determined satisfactionally. Its object is that which satisfies the individual concept expressed by the description 'the F', namely the F. Accordingly, all tokens of that belief-type, i.e., all belief-tokens that the F is G, are about the same object.

Notice that it makes no sense to say that the individual concept *the F* can pick out different objects in different contexts, in which case, *per impossible*, different tokens of the same descriptive belief-type would be about different objects. For if there are many F's, then *the F* is not uniquely satisfied in any context. If a particular belief-token is descriptive and is about a certain F, its content is expressed by a sentence not of the form 'The F is G' but 'The F which is H is G'.[15] Otherwise it is not descriptive but indexical, such that its object is determined contextually.[16]

A *de re* belief-token of the F that it is G is about the F and represents it as being G. It does not represent the F as being F and does not have the F as its object because that object is the F. The object of a *de re* belief-token is that which stands to it in the relation determined by the mode of presentation included in its content. If there is no such object, the belief-token is about nothing at all. However, other tokens of the same type might have objects, and different ones can have different objects.

(B) We can think of the content of a *de re* belief as including, by analogy with indexical expressions in language, indexical elements. Any mode of presentation not expressible by a definite description functions as a mental indexical in this sense. Unlike an individual concept expressible by a definite description, which applies (if to anything at all) to the same object in all contexts, which object is picked out by the mode of presentation in a *de re* belief-token depends on the context of that belief. So we could borrow a phrase from Reichenbach and call these nondescriptive modes of presentation 'token reflexives'.

What is the form of a *de re* belief? Suppose the belief-token is about a certain object *a* under mode of presentation M and it ascribes the property of being G to that object, which in the context M determines to be *a*. A different token of the same belief-type might be about a certain other object *b*, presented in its context by M, and a third token of the same belief-type might be about nothing at all, if in its context there is no object standing to it in the relation determined by M. The first belief-token is true iff *a* is G, the second is true iff *b* is G, and the third, not having an object, seems to have no determinate truth-condition. Yet all three *de re* belief-tokens have the same content, whose elements include M and the concept of G. What is a good way to represent that fact?

Predicates or open sentences won't do, for they do not give complete contents. Yet Burge seems to think they suffice, when he contrasts *de re* with *de dicto* (descriptive) beliefs, which have propositions as their contents, hence context-independent truth-conditions: '*De re* locutions are about predication broadly conceived. They describe a relation between open sentences (or what they express) and objects' (1977, p. 343). However, if *de re* locutions are to give complete contents of *de re* beliefs, the relation they describe must be included in the specification of the content. Burge himself recognizes that there can be more than one kind of *de re* relation (p. 361), but if this is so, then the operative relation must be represented in the content of the belief, so as to determine how its predicative content is applied to an object. Otherwise, there would be no difference in content, as surely there is, between believing of a certain object one is looking at that it is G and believing of a certain object one is recalling, which happens to be the same object, that it is G; and nothing in their contents could determine the relation that contextually determines their objects. Accordingly, the contents of *de re* beliefs are not expressible simply by predicates or open sentences which are contextually applied to objects. Rather, a representation of these contents must include an indexical element expressing the relation which, in the context of thought, determines the object (if any) the thought is about.

Since the truth-condition of a *de re* belief is context-dependent, its content cannot be a proposition, as expressed by a closed sentence. However, to specify its content merely with an open sentence is not to specify enough, as we have just seen. Now an open sentence of the form 'x is G' can be rendered closed either by binding the unbound variable or by substituting for that variable a singular term whose denotation is context-independent. But neither way will do here. The content of a *de re* belief can hardly be expressed by a sentence of the form '$(\exists x)$ $(x$ is G$)$' or even '$(\exists! x)$ $(x$ is G$)$', and we have seen under (2) above that the content of a *de re* belief cannot be a singular proposition expressible by a sentence of the form 'a is G'. Somehow we must include the mode of presentation M in our specification of the content of a *de re* belief, for it determines the relation that contextually determines the object of the belief. Obviously, such a specification could not take the form 'M is G', since the property of being G is not being ascribed to the mode of presentation. And '$(\exists! x)$ $(x$ is presented by M & x is G$)$' won't do, since different tokens with the same content can have different objects. Besides, when the schematic letters are appropriately replaced, these forms express propositions, and those are not what we want.

Having excluded these various ways of representing *de re* beliefs, I offer the following schema as a relatively perspicuous alternative.

$$(\text{DR}) \quad (\exists! x) \, (Rx(\overline{M_R s}) \, \& \, \overline{Gx})$$

The entire schema represents the truth-condition of s's *de re* belief (if need be, a time parameter 't' could be included), and the portions of (DR) under bars represent the contents of s's belief state. $\overline{M_R}$ is the mode of presentation and $\overline{Gx}$ is the predicative content. That is, s represents as being G the unique object that bears R, the relation determined by M_R, to the belief-state (token) that s is in. Since s could have been in just that state without there having been any such object or with some other object having been in R to s's state, (DR) does not require that s's *de re* belief have any object, much less the object it actually has. (DR) represents the truth-condition of a particular belief-token (at a given time) with the content as indicated. The object of that belief-token, that which must be G if the belief is to be true, is whatever object bears R to that belief-token. Notice that although the relation R is represented in the content of the belief, that an object stands in that relation to the belief-token is not represented in it at all. This fact need not be represented precisely because it is by being in that relation to the belief-token that an object is the object of that belief-token.

(C) The contents of *de re* beliefs are incompletely representational. This is suggested by our characterization of their form by means of a kind of open sentence with an indexical expressing a mode of presentation in the position of singular term. However, in calling them incompletely representational I mean something more. Not only are their contents not complete representations (in the same sense in which concepts, as expressed by predicates, are not complete representations), they represent their objects as having certain properties without containing any element that identifies the object being represented. Whereas the identity of the object of a descriptive belief that the F is G is internally represented by the concept *the F*, the identity of the object of a *de re* belief is not represented at all. That is why Burge likes to speak of *de re* belief-contents as being (contextually) 'applied' to an object. If the identity of its object were represented, if the object were thought of in terms of some distinguishing feature, the belief would be descriptive, not *de re*. And there is no way that the 'particularity' or the 'haecceity' (to mention two recently revived medieval notions) of an object can be represented, if indeed objects had particularity or haecceity. Since different tokens of the same *de re* belief-type can be about different objects, there can be nothing in the content of any one such token that makes it have the object it has. Just as there is nothing intrinsic to a picture that makes it a picture of a particular individual (its intrinsic nature determines[17] not what its object is but only what its object is represented as), so there is nothing intrinsic to the content of a *de re* belief that determines which object it is about.

Here I have presented a general conception of *de re* beliefs, not a general theory of them (or of other *de re* attitudes and episodes). Such a theory would characterize the class of *de re* relations in a way that would tell us what they are and how they are determined by different types of mode of presentation. Below I take up the fundamental case of perceptual belief, where the relation is the kind of causal relation essential to perceiving an object and the mode of presentation is a percept. My conjecture is, as regards physical objects (and events) other than oneself, one can have *de re* beliefs not only about objects one is presently perceiving but also about (a) objects one has previously perceived and about (b) objects of others' *de re* beliefs, provided one has appropriate *de re* beliefs about those others and their *de re* beliefs. Unfortunately, I do not have a theory of the contents of and constraints on such beliefs. I suspect that something like the notion of causal chains associated with the so-called New Theory of Reference is needed here.[18] I have no idea what could be the alternative to causal relations and causal chains, and thus take the theoretical problem to be how to characterize them (and how they are determined by different types of mode of presentation).[19] Of course, there are two kinds of *de re* belief which seem to require a different sort of account, beliefs about oneself and beliefs about abstract objects.

III. Perceptual Belief

Surely if we have *de re* beliefs about physical objects other than ourselves, they are about objects we are currently perceiving. And yet Stephen Schiffer (1978) has denied that even perceptual beliefs are *de re* ('irreducibly *de re*') and claims that they are really descriptive in content. First he argues that the content of a perceptual belief cannot be a singular proposition of the form 'F*a*'. Suppose a person *s* perceived the same object twice at the same time without realizing it. Say he is looking at one part of an object *a* and feeling another part, and suppose it looks rough and feels smooth. If *s* believed accordingly, then if the contents of his beliefs were singular propositions, he would believe simultaneously that *a* is rough and that *a* is smooth. Yet surely *s* is not guilty of an outright contradiction—he does not believe that *a* is rough and smooth. Rather, he is merely the victim of ignorance; and surely what he is ignorant of is not that *a* = *a*. So the above singular propositions cannot be the contents of *s*'s two beliefs. Schiffer forcefully argues that the only way to specify the contents of *s*'s beliefs while accounting for *s*'s ignorance is to include in that specification the different modes of presentation under which *s* believes, respectively, *a* to be rough and *a* to be smooth. So far, so good, but what is the requisite type of mode of presentation?

Schiffer proceeds to argue that the only viable candidate for mode of presentation in a perceptual belief is an individual concept as expressed by a definite description (actually an indexical description) containing 'I' and 'now'. Accordingly, if one is perceiving a certain object, one cannot believe something of it *simpliciter* but only under some description. In my view, although Schiffer is correct in arguing that modes of presentation must be included in the contents of *de re* beliefs, it does not follow that these must be individual concepts or descriptions. In particular, it is psychologically implausible to suppose that concepts or descriptions fill the bill in the case of perceptual beliefs. I will suggest instead that percepts are adequate to the task and argue that if they are, then perceptual beliefs are irreducibly *de re*, contrary to Schiffer's descriptive theory.

Appearances and descriptions

Schiffer defies the *de re* theorist 'to specify a counter-example, in which an object of perception is believed by one to be such and such, but where one has no knowledge by description of the object' (1978, p. 195). Now one cannot have knowledge by description of something if the description under which one purportedly thinks of it is not true of it. Indeed, it makes no sense to speak of thinking of an object under a description if that description is not true of it (at best this could mean that one is thinking of the object in some other way, perhaps under some other description, and believes the object thought of in that way to have the property expressed by the description in question). So consider the various descriptions under which one might think of an object one is perceiving or which one might at least believe to be satisfied by that object. Suppose I am looking at a certain object, which I mistakenly take to be a cup, and believe it to be red. I cannot believe it to be red under the individual concept *the cup that I now see before me*. But there is always a fallback description like *the only object which I am now looking at which appears to me to be a cup* under which, according to Schiffer, I believe the object to be red. He maintains not only that plenty of such descriptions are available but that people think of a perceived object they have beliefs about under as many such descriptions as there are available. Accordingly, they have as many beliefs about the object as descriptions under which they think of it.

Schiffer makes no effort to show how it could be plausible psychologically to claim that all these descriptions are not only available to fall back on but that one actually thinks of the perceived object under all of them. Fortunately, Schiffer's descriptive theory of perceptual belief requires something less extravagant, namely that there be at least one such description under which a person thinks of the object of his perceptual belief. In the above example, there must be some fallback description φ, not necessarily the one mentioned above, under which one actually thinks of the object (not merely could think of the object), which description specifies some way the object appears to one at the time.

Schiffer gives no reason why there must be even one such description, some φ under which one actually thinks of the object. If an object appears to me completely different from the way it is but it does not occur to me that the object even might be at all different from how it appears, I am very unlikely to think of the object under any such φ. For example, if I take the object to be a cup and the possibility that it is not does not even occur to me, I will not think of it as the only object which I am now looking at which appears to me to be a cup; I could, but I don't. And so for other such descriptions. Schiffer rightly points out that I cannot doubt of the object, thought of under such a fallback description φ, that it is φ, but not being able to doubt does not entail actually believing. If at the time of the perception I do not think of the object under φ at all, I neither believe nor doubt that it is φ.[20]

In addition to its psychological implausibility, there is an epistemologically unwelcome consequence of Schiffer's overintellectualized position. Suppose an object appears in no way as it is. Then anyone lacking the conceptual resources to think of objects under descriptions like 'the only object which I am now looking at which appears to me to be a cup' cannot have a thought about the object. But surely such a person could, and surely not everyone capable of perceptual belief has the concepts of appearing and of self. It might as well be said that everyone capable of belief must have the concept of truth, as if believing that p requires believing that 'p' (or some sentence meaning p) is true. It is equally implausible to insist that no one can have perceptual beliefs about an object appearing F to him unless he has the conceptual equipment to think of it as the only thing currently appearing F to him.

Now suppose that I do think of the object I am perceiving under a description, not the fallback description Schiffer requires in case the object is not a cup, but simply under the description *this cup*. Whether or not I know the object to be a cup (as Schiffer requires if I am to think of it under that description) and whether or not it is in fact a cup, surely I can think of it as *this cup*. Schiffer's position requires that if it is not a cup, I cannot think of it under that description but must think of it under some fallback description, which it cannot fail to satisfy. But if I can think of it as *this cup* when it is a cup (and I know it to be), in which case I need not think of it under *any* fallback description, then surely I can think of it as *this cup* when it is not a cup. My perceptual experience is just the same, and if I need any description at all in order to think of the object, *this cup* will pick it out, as least for me, just as well if the object appears to be a cup but is not as it would if the object really was a cup. Of course I would not *know* it by description if it is not a cup, but why can I not *think* of it under that description anyway, rather than under some fallback description, if there is no phenomenological difference between the two cases? Only armchair psychology or bad logic—that what I can think I do think—could require that I cannot.

Thinking of a perceived object as *this F* does not require that the object be an F but only that it appears to be. Moreover, contrary to Schiffer's claim 'that "I" and "now" are the only logically proper names that we need to recognize'[21] (1978, pp. 201–2), it is questionable whether 'this', used as a demonstrative pronoun rather than adjective, is eliminable in favour of 'I' and 'now'. Although 'this' as it occurs in 'this is a cup' might be *paraphrased* in some such way as 'the object I am now attending to', it does not seem reducible to that phrase, at least not in a psychologically realistic way, inasmuch as some people, notably small children, can use 'this' without having the concepts of self or of attention. Besides, Schiffer needs to show that 'now' cannot be analysed as 'this moment' and 'I' as 'whoever has this experience or thought'. If some such analysis is correct, then 'I' and 'now' are eliminable in favour of 'this', even if, alternatively, 'this' is eliminable in favour of 'I' and 'now'. If there are equally good ways, from a logical point of view, to eliminate some indexicals in favour of others, which way is to be preferred, if any, is a psychological question.[22]

The contents of perceptual beliefs

There are basically two things wrong with Schiffer's descriptive account of perceptual belief. He has not made a psychologically plausible case that people always think of the objects they perceive under descriptions of the sort his theory requires,[23] and he has not shown that people must think of objects of perception under descriptions of that sort (or any sort) in order to have beliefs about them. Intuitively, the trouble with Schiffer's view is that to believe something of an object one is perceiving does not require thinking of it under any description at all, for it is already singled out for one perceptually. Schiffer is certainly right in rejecting simplistic *de re* theories of perceptual belief, for they are unable to represent the case in which a person believes contradictory things of the same object without being guilty of a contradiction. They cannot represent the possibility that he might not believe, or even disbelieve without being irrational, that each of his two beliefs is about the same object. To allow for this possibility Schiffer follows Frege in positing modes of presentation for the objects of belief, but he does not consider taking them to be, in the case of perceptual belief, percepts rather than individual concepts. This alternative strikes me as eminently plausible not just because the descriptive theory is implausible but because objects of perception are already singled out for us perceptually (Hirsch 1978). The problem is to characterize the contents of perceptual beliefs accordingly.

As noted at the outset, Descartes's epistemologically motivated methodological solipsism has a psychological counterpart. In regard to perception both sorts of MS must make allowances for the possibility of realistic hallucinations. That is, for any putative experience of a physical object, there could be a qualitatively indistinguishable ('realistic') hallucination. That is, one cannot tell merely from having the experience whether or not one is hallucinating. One can tell only by relying on collateral information, but if that information rests ultimately on other perceptual experiences, to which the same point applies, then, so Descartes argued, our perceptual beliefs are not justified at all. Be that as it may, our present concern is not with the justification of perceptual beliefs but with their content. Nevertheless, the possibility of realistic hallucinations requires us to characterize perceptual states in such a way that a person could be in the same perceptual state regardless of whether he is really perceiving a physical object or merely hallucinating. This is precisely what MS demands.

Without committing ourselves to a specific philosophical theory of perception, we can meet this demand by using the 'adverbial' method suggested by Chisholm (1957), whereby types of perceptual states are individuated by the way in which the perceiver is 'appeared to'. This method is ontologically neutral, it avoids commitment to sense data (or anything of the sort), which are posited by representative (indirect realist) and by phenomenalist theories of perception, and it does not depend on direct realist theories either. To be sure, it is most congenial to adverbial (like Chisholm's) rather than act–object versions of direct realism, but that is only because these make the least ontological commitments.[24] The adverbial method of individuation is ontologically noncommittal. It individuates types of perceptual states by their contents, by how the perceiver is appeared to. These contents I call *percepts*. They are not to be construed as what *any* theory proposes as the *objects* ('proper' or 'immediate') of perception. As contents, percepts constitute what perceptual states are like phenomenologically, regardless of the ontological structure of perception.

Let us employ the schema 's is appeared$_m$ to f-ly' to represent a person's being in a certain type of perceptual state (at a time— if need be we could add a time parameter 't'), as determined by the value of 'f' and the sense modality m (visual, tactual, or whatever). We need not restrict, as the classical empiricists would have done, the range of 'f' to purely sensible qualities. It might include properties like being waxen, being rotten, or being an apple.[25] Since percepts are not concepts, there seems to be no definite psychological limit on the complexity of values of 'f'. Phenomenologically there is no clear constraint on the ways in which we can be appeared to, even by physical objects. Since the relevant ways of being appeared to, the relevant values of 'f', are ways that physical objects can appear, I will add a subscript 'x' to 'f' to indicate this restriction. With these qualifications in mind, let us abbreviate our schema as '$A_m f_x s$'.

This schema applies equally to realistic hallucinations and to perceptions of physical objects, and thus using it to represent our way of individuating types of perceptual states is as methodological solipsism allows. The content of a perceptual state, the percept, is given by providing values for 'm' and 'f' (but *not* for 'x') in '$A_m f_x$'. We may represent the state of affairs in which a physical object x actually appears$_m$ f to s as '$Cx(A_m f_x s)$', where 'C' means, roughly, 'causes in the way appropriate to perception' or, more neutrally, 'is related to in the way appropriate to perception'. Just what this relation is, and whether it is causal, is a problem for the theory of perception, not for us, but, with prejudice and simplification, I will paraphrase 'C' as 'causes'.[26] Thus, '$Cx(A_m f_x s)$' says 'x causes s to be appeared$_m$ to f_x-ly' or, more colloquially, 'x appears$_m$ (looks, feels, etc.) f to s'. However, 'appears' should be taken nonepistemically, since there is no implication here that s takes, or is even inclined to take, x to be f. In this regard contrast 'x appears f to s' with 'x appears to s to be f'. In the latter, epistemic sense of 'appears' the property expressed by 'f' must be conceptualized and, since we have placed no restriction on the complexity of values of 'f' in our schema, generally how an object appears nonepistemically includes many more properties than those which epistemically (hence conceptually) it appears to have. For this reason Fred Dretske (1978, p. 124) distinguishes between 'sensory information *available* to the organism' in perception and 'information *actually extracted*' in cognition 'from that which is made available'.

We may now represent the truth-condition of s's belief of the object that appears$_m$ f to him that it is G as follows.

(PB) $(\exists!x)\ (Cx(\overline{A_m f_x s})\ \&\ \overline{Gx})$

The portions of this schema under the bars represent the perceptual and the conceptual contents of s's perceptual belief. Thus, whereas the truth condition of a perceptual belief of the form represented by (PB) requires there to be an object that s is perceiving, since an object must appear$_m$ (f or somehow) to s if s is to have a *de re* belief that it is G, s's psychological state can be specified without implying there to be any such object. Merely the perceptual and the conceptual content of this perceptual belief must be specified to characterize the type of belief-state he is in, and he can be in that state even if he is perceiving nothing at all. In that case there is nothing that s is believing to be G. He believes something to be G only if there is something that appears$_m$ f to him. In other words, the conceptual content of his belief, as expressed by the open sentence 'x is G' applies to an object only if there is an object which is perceptually causing s to be in a perceptual state with content '$A_m f_x$', in which case the belief is about that object. It is in this way, then, that the content of a perceptual belief, like that of any *de re* belief, is not a proposition, expressed by a closed sentence. Rather, its content is expressed by an open sentence with the percept functioning as a mental indexical.

This idea may be clarified by contrasting (PB) with its descriptive counterpart, (PBD):

(PBD) $(\exists!x)\ (\overline{Cx(A_m f_x s)\ \&\ Gx})$

Here the content of the belief and its truth-condition are the same, since the content is a complete proposition. (PBD) explicitly exhibits the two basic difficulties we found with Schiffer's descriptive theory. It overintellectualizes the content of a perceptual belief by requiring, in effect, the believer to conceptualize his perceptual state before he can believe something of its object. In so doing it neglects Dretske's distinction mentioned above. Moreover, it wrongly implies that the individual concept the believer forms from his percept, rather than the percept itself (in the context), determines the object of belief. Rather, the perceiver, by having a percept token appropriately caused by an object, is already in a position to form beliefs about that object. If he first had to form an individual concept (this is implausible phenomenologically in any case), the object of belief would be determined satisfactionally rather than relationally. (PB) represents how the object of a perceptual belief is determined relationally. The percept, expressed by '$A_m f_x$', functions as a mental indexical. The 'reference' of a token of that type is determined relationally; its 'referent' is the object, if any, that bears C to it. To be the object of a perceptual belief, an object need not be represented as being in that relation; it need merely be in that relation. Correlatively, to believe something of an object one is perceiving, one need not represent it as that to which one stands in the relation of perceiving; one is already in that relation to it.

Aside from being much more realistic psychologically than (PBD), (PB) complies with the strictures of methodological solipsism while meeting Schiffer's challenge. Suppose s is perceiving an object twice at the same time but does not realize this. Perhaps he is both seeing it and feeling it; perhaps he is seeing it straight on and in a mirror; perhaps he is 'seeing double'. Let us assume that he is perceiving it in the same modality but that it appears both f and f'. Then according to (PB), s believes of $\imath x Cx(A_m f_x s)$ that it is G and of $\imath x Cx(A_m f'_x s)$ that it is not G. (PB) does not imply or even suggest that s must either believe or disbelieve that $\imath x Cx(A_m f_x s) = \imath x Cx(A_m f'_x s)$. He might have no belief at all on the matter, and (PB) allows that he might even disbelieve the identity. In any event, what s is ignorant of is that the identity holds. Of course, if he did believe the identity, it would be irrational of him to hold both of the perceptual beliefs, for the object cannot be both G and not-G.

Consider now a case in which one object is replaced surreptitiously by another that appears the same. Let us add time indices for before and after the switch, so that (PB) can represent the perceptual beliefs at t_1 and at t_2 as having the following truth-conditions.

at t_1: $(\exists! x) \, (Cx(\overline{A_m f_x s t_1}) \, \& \, \overline{Gx(\text{now})t_1})$

at t_2: $(\exists! x) \, (Cx(\overline{A_m f_x s t_2}) \, \& \, \overline{Gx(\text{now})t_2})$

Here I have made explicit, by including '(now)' in the conceptual content, that s's belief that x is G is not timeless but (present-) tensed, a necessary complication that we have neglected so far. Although the content of s's perceptual belief has not changed (aside from being updated), because of the surreptitious switch the object of the belief at t_2 is different from the object of the belief at t_1. The truth-conditions of his perceptual belief at t_1 and at t_2 are distinct, even allowing for the possibility of different objects, while the content of his belief is unchanged. For he has undergone no change psychologically—his belief-state is the same—and yet semantically his belief has undergone a (Cambridge) change. Suppose the first object has been G all along and the second has been not-G all along. Then s's belief at t_1 is true and at t_2 it is false, despite being the same token of the same psychologically individuated belief-type. Needless to say, there is no violation here of methodological solipsism, for the belief is of an extrapsychological, semantically different type at t_2. Psychologically it is the same.

Perceptual beliefs are about objects without being about the identity of their objects, but of course we have beliefs about that as well. Otherwise, being as short-lived as the perceptual states they involve, perceptual beliefs could not be integrated into our whole system of beliefs. For want of persistent perceptions we can integrate them only by conceptualizing or otherwise transforming perceptual contents. Only thus can the information provided by perceptual beliefs be retained beyond the context of the original perceptions. Although we rejected the descriptive theory of perceptual belief, descriptive beliefs of the form of (PBD) may play a role here, since they are beliefs about objects represented *as* objects of current perception. However, recent work in the field of Artificial Intelligence suggests the need for what A.I. people call nonpropositional representations. At issue here is the so-called frame problem mentioned by Dennett (1978a, p. 125), a latter-day version of Kant's problem about the relation between perception and cognition. The problem is not simply to explain how percepts are put into concepts or otherwise encoded but also to explain how our total system of beliefs is continually revised and, more to the point, updated as we negotiate our way through the world while maintaining a fairly constant model of it regardless of where we are in it. Whatever the cognitive devices we use, surely we must have some way not only of recording perceptual beliefs but of representing the objects of at least some of them as the same individuals as the objects of other beliefs. Perhaps Minsky's 'frames' (1975), or something of the sort, are psychologically more fundamental here than believing descriptively that the objects of certain beliefs are identical.[27] In any case, if we are to maintain a coherent model of the world, we need not merely perceptual *de re* beliefs but also *de re* beliefs about objects previously encountered. It is helpful, though not necessary, to have *de re* beliefs about others and the objects of their *de re* beliefs. However, since their constituent modes of presentation are different, derived *de re* beliefs do not have the same indexical content as that of the perceptual beliefs which anchor them. That is why we are left with the problem of characterizing the types of modes of presentation *de re* beliefs contain and the *de re* relation determined by each type.

Notes to chapter 3

[1] To use Dennett's terminology (this volume), this is the difference between a belief's being 'strongly' and 'weakly about' an object. Note that having an object in mind is not literally having it in the mind. What it does involve is the burden of section II of this paper to explain.

[2] I prefer 'descriptive belief' to '*de dicto*' belief, since the belief is not about the *dictum* but about an object under a description.

[3] Notice that states like regretting, deploring, and rejoicing are not hybrid states in the present sense. Rather, they are complex narrow states that include believing.

[4] Burge (1977) and Schiffer (1977) have both argued that a sentence of the form '*a* is F', where '*a*' is a proper name, cannot express the complete content of a belief. Besides, we have *de re* beliefs about things we do not have names for.

[5] I use 'proposition' without any specific ontological commitment to mean 'content (of a belief or other state) specifiable by a closed sentence'. Whatever the ontological story, this epistemological conception does exclude those uses of 'proposition' that refer to such entities as states of affairs, truth-conditions, and sets of possible worlds.

[6] As with 'proposition' (see note 5) I use 'concept' without specific ontological commitment. Perhaps it will suffice to say that they are those constituents of propositions which represent properties, relations, and, if they are individual concepts (expressible by complete definite descriptions), individuals.

[7] Ambiguity is a property straightforwardly attributed to linguistic (or even pictorial) representations, but it seems inapplicable to internal representations. For any allegedly

ambiguous internal representation would have to be incomplete, in the sense of being meaningful only as part of a 'larger' representation.

[8] My main worry is Burge's admitted restriction of concern to 'our ordinary mentalistic discourse' (1979a, p. 74), as if a determinate commonsense psychology is somehow built into it. The trouble is that the 'that'-clauses in attitude attributions need not be content-clauses, contrary to what Burge seems to assume. Also, he seems to use 'incompletely understands an expression' interchangeably with 'incompletely understands a concept', as if a subject's use of an expression he understands incompletely creates a presumption that the contents of the subject's beliefs asserted with sentences containing that expression include the concept literally meant by that expression rather than some other concept not literally expressed.

[9] A corollary of Burge's position is that 'opaque' does not entail 'narrow'.

[10] In such cases the occurrence of the singular term passes the existential generalization test for transparency but not the substitution test. That the two tests can give different results is one important source of confusion about the transparent/opaque distinction. A more basic source, in my judgement, is the mistaken belief that the distinction belongs to linguistic semantics (as opposed to linguistic pragmatics and psychosemantics).

[11] Rather, the phenomenon is semantic indeterminacy. Claiming semantic ambiguity would commit one to the implausible view that a sentence with n terms each of which could be used either transparently or opaquely is ambiguous in 2^n ways (at least).

[12] Although it is common to refer to 'that'-clauses in belief-ascriptions as content-clauses (e.g., McGinn, this volume, and Schiffer 1980) it is a mistake to do so without qualification, for elements in the 'that'-clause need not express elements of content in the belief ascribed.

[13] It should be noted that Schiffer (1978) appeals to modes of presentation in the course of trying to reduce *de re* beliefs to descriptive ones, whereby they appear, with mode of presentation included, to have propositions (but of course not singular ones) as contents. In discussing perceptual beliefs in section III, I argue that while their contents do include modes of presentation, percepts not descriptions, their contents are still not propositional.

[14] Here and later I use 'predicative content' for what is in superficial predicate position. Of course, if Russell's theory of descriptions is correct, as I believe it to be (Bach forthcoming b), then general terms contained in descriptions appear in predicate position on analysis.

[15] Whereas it is common for syntactically definite descriptions (like 'the book') to be used with implicit qualifiers (like 'which I am reading'), usually further descriptive expressions or indexicals, for the addressee to infer (see Bach forthcoming b), the analogous point cannot be made for contents of beliefs. There is no distinction in thought corresponding to the 'explicit/implicit' distinction in language use between what a speaker says and whatever else he means but leaves to be inferred. Of course, there is another distinction in thought, that between explicit and implicit representation, but I assume that implicit representations are included in contents of thoughts.

[16] I will use the term 'indexical' broadly to refer to descriptions containing indexicals (pronouns) in the narrow sense, as well as to beliefs whose contents are fully expressible with indexicals in subject position.

Note that not every indexical belief is *de re* with respect to its object. For example, my belief that the window behind me is presently open might be *de re* with respect to myself and to the present moment but not to the window. The object (which window is) being thought about is determined descriptively, given the identities of the thinker and the time of the thought.

¹⁷ The determination is relative to a system of pictorial representation. See Goodman (1968) and Bach (1970).

¹⁸ In my opinion what is right about the New Theory of Reference applies not to (linguistic) reference at all but to the determination of the objects of *de re* beliefs. The theory has been motivated chiefly by the claim that names are rigid designators, but I have argued for a descriptive theory of names that refutes this claim, at least if construed as about the semantics of names (Bach forthcoming a). And Schiffer (1977) has shown that in belief-ascriptions content-clauses with names in subject position are incomplete for not specifying modes of presentation. So if the notion of causal chains is to be used for a theory of *de re* belief rather than for the theory of linguistic reference, it will apply to modes of presentation instead of names or their mental counterparts.

¹⁹ Schiffer (1978 pp. 186–9) rightly rejects causal chains as modes of presentation, since they cannot be parts of belief-contents, but he does not consider the possibility that a mode of presentation could *determine* a causal chain. Also, as Mike Harnish has pointed out to me, Schiffer seems to assume that a causal chain can be specified only by specifying the object to which it extends.

²⁰ I am not assuming here that if it does not occur and has never occurred to one that *p*, then one does not believe that *p*. On the other hand, I am not assuming, as Schiffer seems to do, that for any *p* which a person is capable of formulating, he either believes that *p* or withholds believing that *p*.

²¹ Schiffer is using 'logically proper name' in the way that Russell did, even though pronouns do not denote semantically, i.e., independently of context.

²² In their efforts to analyse both *de re* and descriptive beliefs in terms of self-ascriptive beliefs, Lewis (1979) and Chisholm (1980) both ignore the question of psychological plausibility.

²³ As I have urged in regard to the claim that intention is essential to action (Bach 1978), when philosophical analyses of mental notions become too complex to be plausible psychologically, it is time to weaken one's analyses or appeal to weaker notions.

²⁴ This does not make them right, any more than the claim that nothing exists. My main complaint about adverbial theories is their difficulty in accounting for the objectival character of experience, especially visual and tactual, without introducing the notion of intentional objects. To avoid this difficulty I once proposed (Bach 1968) the 'perceptual object theory', on which all experiences have objects, either physical objects or, in the case of realistic hallucinations, 'private' objects.

²⁵ If types of objects are included, perhaps they should be restricted to those that have characteristic ways of appearing.

²⁶ If the relevant relation is relative to the sense modality, we could easily add a subscript to yield 'C_m'.

²⁷ Perhaps these are the means by which we represent the 'notional objects' which, according to Dennett (this volume), comprise our respective 'notional worlds'. However, since we 'live in the same real world', at least some of our respective notional objects must be realized by the same physical objects, and for this perceptual *de re* beliefs, including plenty of true ones, are indispensable. This is the main reason, I think, that people have complained that present-day, artificially intelligent beings lack *Dasein* (to borrow a term from Jerry Heidegger).

4

On the Ascription of Content

Stephen P. Stich

The former Prime Minister of Sweden, Mr Fälldin, believes that nuclear power plants are unsafe. His belief that nuclear power plants are unsafe led to the fall of his government.

In saying what I have just said, I have attributed a belief to Mr Fälldin and gone on to comment on some of the consequences of that belief. The question I want to explore in this paper is how I did it. The beginning of the story is obvious enough. I attributed the belief by producing a sentence of the form:

S believes that p,

and I commented on its consequences by producing a sentence of the form:

S's belief that p ——.

Sentences of these forms are both common and flexible. By varying the sentence that takes the place of 'p' I can attribute to S any of an indefinitely large number of beliefs. And therein lies a puzzle. For in using sentences of the form *S believes that p*, I am saying that S has a belief—that he is in a certain kind of psychological state. And it is the 'content sentence', the sentence replacing 'p', which serves to indicate *which* belief it is. But how is this possible? What is the relation between the psychological state and the content sentence which enables me to use the latter to identify the former?

I. Assumptions

To sharpen the focus of my question and to lay a foundation for my answer, I will help myself to a healthy serving of assumptions. Some of these assumptions have been widely discussed and defended in the philosophical literature. Others I have myself defended elsewhere. And a few, I am afraid, are no more than promissory notes.

My first assumption is that we all share a (largely) tacit theory according to which the behaviour of people and higher animals is to be explained (in part at least) by appeal to their beliefs and desires. It is in virtue of being embedded in this 'folk psychology' that such terms as 'belief' and 'desire' acquire their meaning. My project in this paper is to make a bit of this folk psychology explicit. Thus, like many projects in analytic philosophy, I will be engaged in a sort of anthropology, a study of folk beliefs and concepts where we ourselves are the folk in question. The study of folk beliefs or conceptual systems is not the exclusive province of analytic philosophy and anthropology. When done on a finer scale it is the project of the cognitive simulator, the psychologist who seeks to make our tacit beliefs explicit in sufficient detail to simulate them with a computer program. It is my view that philosophical analysis, when done well, is continuous with the project of the cognitive simulator. The philosophical analyst can be viewed as giving a discursive characterization, a sort of sketchy flow chart, for the program the simulator is trying to write. This is how I would have the present project construed.[1]

Our tacit theories, be they folk psychology or folk physics or what have you, manifest themselves in various ways in our everyday behaviour. Among the manifestations that loom large for the simulator are the judgements we make about the application of a term to various actual and possible cases. These judgements, though of no great intrinsic interest, allow the theorist to study the workings of one or another aspect of the subject's conceptual system in *relative* isolation from the rest of his beliefs, concepts, motivations and skills. So as a strategy for uncovering our tacit theory of belief I propose that we think of ourselves as trying to write a program that will match our intuitive judgements about when it is, and is not, acceptable to say of a given person that he believes that p. Given a description of a case, such a program should be able to render a judgement on what beliefs are appropriately attributable to the protagonists, and that judgement should match the one we ourselves would give. I hasten to add that the goal of actually producing such a program is a *distant* goal. And, though I hope the present paper may throw some light on what parts of a simulator's program might look like, I have made no systematic effort to show how the simulator's project might join up with the analysis I offer.

Beliefs are psychological states, and it is inevitable that the story I have to tell will involve much talk about states and their types. So I had better say something about how I would have such talk construed. I shall take states to be the instantiation of a *property* by an *object* during a *time interval*. So construed, states are particulars with more or less definite locations in space and time. I will use expressions of the form '(O, P, Δt)' to denote the state which is the instantiation of property P by object O during time interval Δt. I will also sometimes use expressions of this form to talk about *possible* states. My claims about possible states can be taken as shorthand for counterfactual claims. Thus

Possible state (O, P, Δt) is ϕ

should be construed as

Were O to instantiate P during Δt, the resulting state would be ϕ.

On the account of states that I shall adopt, states admit of what might be called an *essential* classification into types. A pair of states are of the same *essential* type if and only if they are instantiations of the same property.

It will also sometimes be convenient to use the word 'state' to denote the property which all states of the same essential type share. Thus, when I say of a certain object that it is *in state P*, I will mean that the object instantiates property P at the time in question. Similarly, to say that a pair of objects are *in the same state* is to say that they instantiate the same essential property at the times in question. When ambiguity threatens, I will use 'state token' to refer to particulars and 'state type' to refer to properties.[2]

Although each state token has only one essential type, states, like other particulars, can be grouped into non-essential types in an endless variety of ways. A type of state tokens is simply a category of particulars, and we have specified such a type when we have set out conditions for membership in the category. Similarly, state types, both essential and non-essential, can themselves be grouped into types or categories. To specify a category of state types we need but specify the conditions under which a type will count as a member of the category.

My next assumption is one that would have been roundly dismissed in philosophical circles twenty years ago. But by now it has become quite an orthodox view. This is the claim that folk psychological explanations of behaviour are to be understood as ordinary *causal* explanations. Thus, when we say, 'Mr Fälldin prohibited the activation of several nuclear power plants because he believed that they were unsafe', we are saying that the former Prime Minister's belief was one of the causes of his action.[3]

Our folk psychology has a fair amount to say about the 'functional' properties of the states and mechanisms it deals with, i.e. about the causal interactions of these states and mechanisms with one another, with stimuli and with behaviour. However, it says little or nothing about the physiological nature of these states. Folk psychology is not only physiologically modest, it is psychologically modest as well. There is no assumption that the causally or functionally specified features of the states postulated are the *only* causal features of these states that are psychologically relevant. Folk psychology is prepared to accept a wide range of empirical discoveries about factors affecting the formation of beliefs and desires, and about their interaction with other states or with behaviour.

A central tenet of our folk psychology is that there are two quite different kinds of psychological states, beliefs and desires, which, along with other sorts of states, interact to produce behaviour. Normal subjects have large numbers of beliefs and large numbers of desires, some fleeting and some enduring. Folk theory has a fair amount to say about the typical causes and effects of beliefs and desires. Beliefs, for example, can be caused in a number of ways, the two most conspicuous being through perception and through inference. Inference, in turn, is a process whereby beliefs interact with one another and give rise to other beliefs. Desires, by contrast, are often caused by deprivation. A person deprived of food, drink or sex, to mention the three most obvious examples, will typically acquire a desire for food, drink or sex, with the strength of the desire roughly correlated to the length of deprivation. Desires can also be caused by the interaction of beliefs with other desires. Thus my desire to see Solti conduct, along with my belief that he will be conducting in London next Saturday, led to my forming the desire to go to London next Saturday. And this desire, along with my belief that to get to London I must buy a train ticket, led to the desire to buy a ticket. At the end of these chains of practical reasoning are basic desires about bodily movements and the like, which are capable of causing behaviour.[4]

Amongst people, at least, where our folk psychology is most at home, the theory recognizes a distinction between those beliefs that a person is conscious of holding at a given time and those that he is not conscious of holding. The vast majority of our beliefs fall into the second category—not-currently-conscious beliefs. This category should not be confused with the category of subconscious beliefs, the latter being a subset of the not-currently-conscious beliefs whose members, for one reason or another, *cannot* be brought to consciousness. It is likely that in pre-Freudian times folk psychology held that people could become aware of all their not-currently-conscious beliefs, but by now the notion of subconscious beliefs (and desires) has become well entrenched in common-sense theory. The distinction between currently conscious and not-currently-conscious beliefs does not coincide with the distinction between beliefs which are currently being *used* (in inference, practical reasoning, etc.) and those which are not currently being made use of. It is often the case that we make use of our beliefs without being aware that we are doing so. The most blatant example is the role that subconscious beliefs may play in inference and practical reasoning, and thus in the determination of behaviour. But even those beliefs which can be brought to consciousness often are psychologically active without our being aware of it. Thus it is common to hear a person report that he solved some problem without even being aware that he was thinking about it.[5]

Our intuitive theory recognizes a distinction between the beliefs a person has at a given time and the beliefs he might quickly come to acquire by inference, should the need arise. So it generally makes good sense to ask such questions as: 'Did the inspector believe that George was the murderer as soon as he heard the details of the case, or did he come to that view only after thinking it through?' And students facing tough competitive examinations are sometimes advised that it is better to have as many formulas as possible committed to memory, rather than having to figure them out when they are needed. However, in many cases we infer beliefs very quickly, and without being conscious that we are doing so. In these cases it is difficult or impossible for a person to know whether the belief in question is one he had previously or one he inferred on the spot when his attention was directed to the matter. To make things worse, common usage is often quite tolerant about belief-attributions, and thus it is not at all unnatural to ascribe to a person almost any belief that he would quickly and unconsciously infer from the beliefs he already has. So it would not be counter-intuitive to assert that Bertrand Russell believed (indeed knew) that Big Ben was larger than Frege's left ear lobe, even if, as is almost certainly the case, Russell would have had to infer that belief from other beliefs, were the question ever to have arisen. Still, I think our intuitive belief–desire psychology is congenial to the proposal that psychologists might discover ways of distinguishing beliefs we already have from those we quickly and unconsciously infer when they are needed. Thus our folk theory can comfortably concede that perhaps Russell didn't actually believe Big Ben is larger than Frege's ear lobe, though he certainly would have come to believe it, had he been asked.[6]

Folk psychology does not take individual beliefs to be simple and unstructured. Rather, beliefs are composite states whose various parts can recombine in different ways to form different beliefs. The term *concept* is sometimes used to denote one sort of constituent of beliefs. Thus Jones's belief that Abraham Lincoln was born in Illinois might be said to be composed of Jones's concept of Lincoln, his concept of Illinois, his concept of birth, etc. In order to have a given belief, it is necessary that a person should have the appropriate concepts. So to believe that genes are made of DNA one must have the concept of a gene and the concept of DNA. There is a conspicuous correlation between the concepts we would ordinarily say are involved in a belief and the *words* that we would use to express the content of the belief. Some theorists have urged that beliefs might plausibly be *identified* with sentences (or sentence-like entities, such as phrase structure trees) which are encoded and stored in the brain. Concepts could then be identified with words in the language of thought.[7] I have considerable sympathy for this proposal as a paradigm for theorizing in cognitive psychology (but see also Stich 1978b, 1980, forthcoming; and section VII below). Though, as most advocates of the view would cheerfully concede, folk theory demands no such sententialist account of beliefs. For present purposes I will assume only that common-sense psychology takes beliefs to be structured states (some of) whose components are concepts, and that the concepts composing a particular belief are at least roughly correlated with the words we would use to express the contents of the belief. It would be interesting to explore how much more detail could plausibly be attributed to the concept concept in folk psychology. But I will not attempt the task here.

These brief remarks are intended as no more than a gesture at the quite complex set of common-sense principles that specify the ways in which beliefs and desires may causally interact with each other, with other categories of psychological states and with behaviour. Those principles taken together serve to characterize what might be described as the *global architecture* of our folk psychological theory. The story I want to tell about the ascription of content assumes very little about the details of this global architecture. However, I will suppose that the global architecture of our folk theory is sufficiently detailed to divide the complex systems of the world into two categories, those to which it applies and those to which it does not. Those to which it applies will have to have sub-systems of physical or physiological states whose causal interactions with each other, with environmental stimuli and with the system's behaviour are as specified by the global architecture of our intuitive psychological theory. For the complex systems to which it applies, the global architecture of our folk theory will have to distinguish a category of physical or physiological states which behave as belief-states within the system, and another category of physical or physiological states which function as desire-states within the system. Note that the categories of belief-states and desire-states are what I earlier called non-essential types of states. Each particular belief-state or desire-state is also a physical or physiological state, and its essential type is determined by some physical or physiological property.

A complete account of the global architecture of our intuitive psychological theory would be far from an exhaustive specification of our folk psychology. For, in addition to these general principles about the kinds of causal interactions into which beliefs and desires enter, we also have a wealth of more detailed theory. Consider inference, for example. Our informal theory tells us much more than the mere fact that inference can lead to the creation of new beliefs and the elimination of old ones. It also tells us a great deal about which beliefs are likely to be generated from which. If Sam believes that terrorists killed everyone aboard the flight to Fort Lamy, and if he then comes to believe that the President of Mali was aboard that flight, he will likely come to believe that the President of Mali is dead. It is much less likely that these two beliefs will cause him to believe that more than half the people of Albania are Muslims. I will call this detailed theory about which beliefs and desires are likely to be caused by which, which stimuli are likely to lead to which beliefs, etc., the *fine structure* of our folk psychology. The numerous principles that make up the fine structure of our intuitive psychology, unlike the principles of gross architecture, are all cast in terms of the content of beliefs and desires. That is, in detailing which beliefs are likely to be generated by which, the beliefs are identified by specifying their contents. Our present project is to explain how it is possible to identify a belief by specifying its content. It thus constitutes a sort of prolegomenon to any account of the fine structure of folk psychology.

II. The logical perversity of belief-sentences

From Frege's time onward, philosophers have devoted much attention to the peculiar logical behaviour of sentences of the form *S believes that p.* In logically well behaved contexts, co-designating names and co-extensive predicates can be substituted for one another without risk of changing the truth-value of the sentence containing them. And even in some logically not so well behaved contexts, a sentence embedded within another sentence can be replaced by a logically equivalent sentence without risk of changing the truth-value of the embedding sentence. But none of this is true for substitutions which tamper with the embedded sentence following 'S believes that ——'. There any of the substitutions mentioned might potentially change a true belief-sentence into a false one. To make matters worse, many philosophers have urged that 'believes' is ambiguous and thus that belief-sentences have *two* quite distinct senses, only one of which exhibits the full range of logical perversities I have described. The other sense, though still obdurate about substituting co-extensive predicates or logically equivalent content sentences, permits the free substitution of co-designating names and definite descriptions, *salva veritate.* Philosophers who endorse the view that belief-sentences are thus ambiguous have labelled the first sense *de dicto* or *opaque,* and the second *de re* or *transparent* (cf. Quine 1956; Kaplan 1968).

Now it is my view that the alleged ambiguity of 'believes' is actually an illusion. I try to make the case against the *de dicto/de re* distinction in a paper (in preparation) designed as a sequel to the present one. Since I reject the view that belief-sentences are systematically ambiguous I intend the account I give of belief-sentences in the following pages to apply equally to all belief-sentences, not just to one or the other alleged sense of such sentences. Still, I would hope that the account I give of content ascription might be of interest even to those philosophers who reject my rejection of the *de dicto/de re* distinction. Readers who are partial to the distinction are urged to view the current essay as a study of content ascription in *de dicto* belief-sentences only.

III. The beginnings of an analysis

In the present section I want to begin work on an analysis of belief-sentences which will explain how these sentences can be used to identify and attribute psychological states. The account

I will elaborate is rather a patchwork product, pieced together in part from the hints and ideas of a number of other philosophers.

To start, let me recall a few of the moves in Carnap's evolving analysis of belief-sentences. As a first approximation, Carnap (1956, section 13) proposed we might analyse

(1) John believes that snow is white

as

> John is disposed to respond affirmatively to 'Snow is white' or to some sentence which is L-equivalent to 'Snow is white'.

Here, 'respond affirmatively' means something like 'would agree if asked', and L-equivalence is a logical equivalence relation that can obtain between sentences in different languages as well as between sentences in the same language. The last clause in the analysans is included to cover the case in which John does not speak English. Carnap quickly realized that the analysis is faulty on at least two counts. First, L-equivalence is not a strong enough relation to do the needed job in the last clause of the analysans. For as it stands the analysis entails that if a person believes that p, and if q is (however non-obviously) logically equivalent to p, then the person believes that q as well. But as we noted in the previous section, this is not the way belief-sentences behave. The second problem Carnap finds with his preliminary analysis is that the connection demanded between belief and behaviour is much too stringent. It is easy to describe cases, real and imagined, in which a person plainly believes that p but, because of his other beliefs and desires, is not at all disposed to respond affirmatively to p, nor to any other sentence in any language that is L-equivalent to p. In an effort to handle the first problem, Carnap (ibid., section 14) experimented with a tighter equivalence relation than L-equivalence, which he called 'intensional isomorphism' (cf. Davidson 1963b). To handle the second problem, Carnap proposed that the term 'believes' be treated as a theoretical term in a psychological theory. In the context of that theory, a sentence like (1) 'can neither be translated into a sentence of the language of observables nor deduced from such sentences' (ibid., p. 230). At best, (1) can be inferred 'with high probability' from an observation report like

> John makes an affirmative response to 'Snow is white' as an English sentence.

There are two central ideas in Carnap's account that will be knitted into my own. The first is the suggestion that, despite the appearances of surface grammar, 'believes' be treated as a predicate expressing a relation between the believer and some linguistic object. On Carnap's view it is the content sentence itself, though that is a detail I shall ultimately reject. Carnap would have us take the logical form of (1) to be something like

> (2) Believes (John, 'Snow is white').

An obvious attraction of this move is the ready explanation it offers for the logical perversity of belief-sentences. For on this account of the logical form of belief-sentences, the content sentence is not used in the analysans, it is mentioned. Substitutions of co-designating terms, co-extensive predicates or logically equivalent content sentences are thus substitutions within a quoted context, and we would not expect them to preserve truth-value. The second Carnapian idea that I will borrow is the suggestion that 'believes' be treated as a *theoretical term* embedded in a psychological theory. Unfortunately, Carnap tells us very little about the theory in which the belief-predicate is embedded, nor does he tell us how the belief-predicate integrates with the rest of the theory. For our purposes, however, the *choice* of a theory is obvious. The task we have set ourselves is to do a bit of domestic anthropology, to understand what we are doing when we use sentences like (1) in everyday discourse. And if we are making a 'theoretical' claim about John when we say he believes that snow is white, then surely the theory is our own intuitive folk psychology. So let us reflect on just where the cogs of our folk psychology link up with sentences like (1).

As a first step, recall our assumption that the gross architecture of our folk psychology characterizes a category of physical or physiological states which function as belief-states for the system or organism in question. Now when we say John believes that snow is white, surely we are saying that he is in some state functioning as a *belief*-state. So let us build this notion of belief-state into our analysis, replacing Carnap's (2) with

(3) $(\exists b)$[John is in b & (John, b, t_n) is a belief-state & B((John, b, t_n), 'Snow is white')].

Here b must be a state type (i.e. a property), presumably a physical or physiological property, and t_n is some time interval containing the present. I have traded Carnap's unexplicated 'Believes' relation (which obtained between a believer and a content sentence) for an equally unexplicated relation, B, between a belief-state and a content sentence. B is that relation, whatever it may be, which enables us to use a content sentence in identifying the belief-state we are attributing. This is scant progress, however, since it remains to ask just what this relation B is.

Well, how is a content sentence related to the belief-state it serves to identify? Here, as earlier, I think some of Carnap's remarks provide an important insight along the way to an answer. According to Carnap,

> a sentence like 'John believes that the earth is round' is to be interpreted in such a way that it can be inferred from a suitable sentence describing John's behavior at best with probability, but not with certainty, e.g. from 'John makes an affirmative response to "the earth is round" as an English sentence' (1956).

Now, putting quibbles to one side, there can be little doubt that a considerable part of our inference about our co-linguists' beliefs follows the pattern Carnap describes: from assent to or assertion of the sentence p to the belief that p.[8] But why is this pattern of inference generally reliable?

The answer I would urge here is a variation on a theme developed by David Lewis (1966, 1970, 1972b) in his account of common-sense psychological states like pain. Consider an analogy. Why is it that 'pain behaviour' is good evidence that the subject exhibiting the behaviour is in pain? Lewis's answer is that pain simply *is* that psychological state which typically causes pain behaviour. A bit more precisely, Lewis suggests that we characterize psychological states by specifying their *typical causes* and/or their *typical effects*. Thus, for example, on Lewis's view pain would be characterized as that psychological state which was typically caused by damage to the skin, etc., and typically causes various sorts of 'pain behaviour', etc. Since the characterization is in terms of typical causes and effects, it is left as an open possibility that there may be atypical cases of pain which cause no pain behaviour, and atypical cases of pain behaviour which are not caused by pain. Now on my view our inference from the assertion of p to the belief that p is quite parallel to our inference from 'pain behaviour' to pain. In both cases we are inferring from an effect to its typical cause. And in both cases the cause, the underlying psychological state, is characterized in terms of its typical effect. So, as a first pass, the view I want to defend is this: the belief that p simply *is* the belief that typically causes the assertion of p. The relation B, then, is the relation of typically causing, and in using locutions of the form *S believes that p* we are identifying a belief-state by citing the sentence whose utterance it typically causes. But this fast and dirty adaptation of Lewis's idea to the case of belief is infested with problems. In the remainder of this section, and all of the section that follows, I will explore one strategy for constructing an acceptable Lewis-style theory about beliefs.

A first problem with taking B in (3) to be the relation of typically causing is that John may not be an English speaker. And if he is not, then surely *his* belief that snow is white does not typically cause utterances of 'Snow is white'. So we had better build this recognition of the parochialness of belief-expression into our account of the B relation. Rather than taking B to be the relation of typically causing, we can view it as the relation of typically causing *chez nous*, amongst our own co-linguists.

When we make this move, however, we create a new problem for our fledgling analysis. Whatever plausibility there may be to the claim that John's belief-state-token typically causes utterances of 'Snow is white' if he is an English speaker, it is preposterous to suggest that *his* belief-state-token typically causes the utterance of a sentence *amongst us* when John is not one of us. Clearly what is needed is an appeal to *types* of beliefs. The idea is that John's belief-state-token is of a type which, *chez nous*, typically causes utterances of 'Snow is white'. This suggests that (3) be replaced by something like

(4) $(\exists b)[$John is in b & (John, b, t_n) is a belief-state & B(b, 'Snow is white')$]$

with B construed now as the relation holding between a property (i.e. a state type) and a sentence when instantiation of that property *amongst English speakers* typically causes the utterance of that sentence.

There was a time when it was the fashion to think that psychological state types could be identified with physical state types, that a toothache (anybody's toothache) simply was the firing of a certain type of neuron. To theorists of that persuasion, (4) might appear quite congenial. Recall that the category of belief-states is a category of physical state tokens. So for (John, b, t_n) to be a belief-state, b must be a physical property. And (4) urges that whenever a subject believes that snow is white he does so in virtue of instantiating a single physical property. Nowadays, however, few philosophers would find this 'type–type' identity theory plausible, since it offhandedly rules out the possibility that robots, Martians and folk with physiology different from our own might believe that show is white. Yet surely it was no part of our intention in shifting from (3) to (4) to embrace a type–type identity theory. The idea behind the move from (3) to (4) was that John's belief that snow is white must be of the same type as the belief which, amongst English speakers, typically causes the utterance of 'Snow is white'. But, as the present reflections show, we cannot construe 'same type' as 'same physical type', on pain of denying that Martians and robots have beliefs. So there must be some other notion of type identity for belief-state-tokens, some notion which can count a pair of belief-state-tokens as type identical even though they are different physical states, and thus not members of the same essential type. What my belief that snow is white, John's belief that snow is white and the robot's belief that snow is white have in common is not their physical characteristics but their content. It looks like what we need is some notion of *content identity*, a relation that can obtain between belief-states even though they are not of the same physical type.

If we had such a notion of content identity available, we could replace (4) with something like the following:

(5) $(\exists b_1)(\exists b_2)$[John is in b_1 & (John, b_1, t_n) is a belief-state & (I, b_2, t_n) is a belief-state & B((I, b_2, t_n), 'Snow is white') & Content Identical ((John, b_1, t_n), (I, b_2, t_n))].

The 'I' in (5) is to be read as an ordinary indexical, referring to the speaker. So what (5) says is that there is a belief-state-token of John's which is content identical with the one which typically causes *my* utterances of 'Snow is white'. Now, as will be evident even to the most sympathetic reader, (5) is not a suitable endpoint for our analysis. But a less sympathetic reader might protest that it is not even an acceptable starting point for further work. 'It is folly', such a reader might protest, 'to invoke a notion of content identity in an account of content ascription; to do so is to beg most of the interesting questions.' The complaint is no doubt a reasonable one, and it signals the major fork in the roads I am exploring. What I propose to do, for the time being, is to simply assume that a suitable notion of content identity is available, and to push on with my attempt to give a Lewis-style account of content ascription. When I have followed that route as far as I can, I will return to the present junction and explore what can be done to cash the concept of content identity in more creditworthy coin.

Let us return, then, to (5) and ask how well it serves as an analysis, assuming some suitable account of content identity is at hand. An obvious problem is that, as it stands, (5) commits the speaker to *holding* all of the beliefs he attributes to others. This is implied by the fourth conjunct which says that a certain belief-state *of the speaker* typically causes his utterance of 'Snow is white'. Another problem, only a bit less obvious, is that (5) seems to require the speaker to be prolix beyond imagining. For suppose that I were now to tell you:

The President believes that The Pope was Archbishop of Cracow.

And suppose further, what is in fact true, that I believe it too. Then, according to (5), what I have told you, roughly, is that there is a belief of the President's which is content identical with the one of mine that typically causes my utterance of 'The Pope was Archbishop of Cracow'. But this cannot be right. As best I can recall, I have never *uttered* the sentence 'The Pope was Archbishop of Cracow'. So if 'typically causes' means *causes most instances of*, then there is *no* belief which typically causes my utterances of 'The Pope was Archbishop of Cracow'.

One way to escape this difficulty is to reinterpret the notion of a typical cause. Rather than invoking a straightforwardly statistical notion of typical cause we can construct a new one built on the idea of a *typical causal pattern*. The basic idea here is that there is a typical or characteristic *sort* of proximate causal history that underlies most of our assertions. Moreover, if we pair our co-linguists' beliefs and desires with our own via the notion of content identity, then we expect that their assertions are, by and large, the product of causal histories fitting the same pattern. This is not, of course, to deny that many of our assertions and those of our co-linguists have proximate casual histories which are both devious and deviant. We sometimes say what we do not believe, and we sometimes say what we do believe only with the most devious of reasons. Yet most of our assertions are sincere expressions of belief (cf. Lewis 1969, ch. 5). Now what I am assuming, and what I think our common-sense psychology assumes, is that there is a *common causal pattern* underlying most cases in which we sincerely express a belief. Given the preponderance of sincere assertion, this common causal pattern is the *typical causal pattern* underlying our assertions. No doubt the pattern, like all patterns, admits of considerable variation. Yet I think we ordinarily suppose that the proximate causal histories underlying sincere assertions are distinguished by some important common features, that they constitute a single kind of psychological process. It is perhaps a symptom of this tenet of our folk psychology that we find the idea of constructing a lie detector well within the bounds of conceptual possibility. Indeed, I am inclined to think that the impressive, though partial, success of the current generation of lie detector technology is some evidence that our folk theory is correct in its assumption.

Now granting the notion of a typical causal pattern leading to assertions, how can the notion be woven into our account of the relation B in (5)? A first thought is that we identify the belief that p as *the* belief which can play a role in a typical causal history leading to an utterance of p. But, as a bit of reflection will reveal, this will not do. The problem is that many beliefs in addition to the belief that p may play a role in a typical causal history leading to a sincere assertion of p. Thus, on one occasion I may assert 'Lead floats on mercury' because I believe it, and because I believe that you have just asked whether lead floats on mercury. On another occasion I may make the same assertion because I believe lead floats on mercury and because I believe you are dangerously misinformed on the matter. However, in all of these cases, I think we assume the belief that p plays a *special central* role in typical causal histories leading to the utterance of p. We might pick out the role as follows. Consider those eternal sentences which have been asserted with considerable frequency. There will generally be only one belief which is part of the causal history of each typically caused utterance. And it is the role which this belief plays that we are calling the *central* role. I rather suspect that in specifying this way of picking out the central role a belief can play in typical causal histories we may be going beyond our common-sense psychology. It may be that folk theory gets by with the mere assumption that *there is* a special central role for the belief that p to play in typical causal histories of the utterance of p, without ever worrying about how this special role might be picked out. In any event, with the notion of a central role in a typical causal history at hand, we can give the following analysis of the B relation:

> A belief-state (I, b, t_n) stands in the relation B to a sentence p if and only if (I, b, t_n) (or a content identical belief-state) could play the central role in a typical causal history of an utterance of p by me (or by one of my co-linguists).

A bit less formally, the idea I am urging is this. In identifying a belief as *the belief that p*, we are picking out the belief by producing an example of the effect it typically has among us, where 'typically' is construed not statistically, but by appeal to a typical causal pattern and a special role within that pattern.

We are, however, still not finished patching our analysis. For even with the newly proposed construal of the B relation, (5) still entails that a speaker holds all the beliefs he attributes to others. I will not pause to patch that problem though, since another, larger, problem looms. The steps taken to deal with it will handle the former problem as well.

IV. On ambiguity

Natural languages are notoriously ambiguous. A given sentence often can be used to express many different beliefs. Also, many sentences contain names or definite descriptions which do not uniquely denote. Both of these facts pose serious problems for our analysis in its current form. To see this consider the following sentence:

Nixon believes that John Dean is a Russian agent.

On a recent inspection of the telephone directories for Washington, D.C. and its suburbs, I found listings for more than a dozen John Deans. And the content sentence

John Dean is a Russian agent

might be used to express an unflattering belief about any one of these Washingtonians. That is, there could be upwards of a dozen quite distinct beliefs, any one of which could play the central role in a typical causal history leading to an utterance of 'John Dean is a Russian agent'. This fact might not pose a serious problem for my account if actual belief-attributions were similarly ambiguous. But the fact is that they are not. To be sure, belief-*sentences* when considered in isolation inherit all the ambiguity of their content sentences. But specific uses of belief-sentences are rarely infected by this ambiguity. Thus, if I tell you now that Quine believes Frege was a major figure in the history of logic, I will succeed in (truly) attributing a single belief to Quine, no matter how many Freges there may be. In general, when a belief-sentence is actually used to attribute a belief to a person, the referring expressions in the content sentence have the denotation they would have were the content sentence alone to have been issued in the identical setting. Similarly, a potentially ambiguous content sentence generally renders the surrounding belief-attribution ambiguous only to the degree that the content sentence would be ambiguous if uttered alone in the same setting. Some strategy must be found to take account of these facts in the analysis of belief-attributions.

One solution to our problem is to borrow an idea due to Davidson. In his account of indirect discourse Davidson (1968) urges that we abandon the Carnapian strategy of analysing sentences like

Galileo said that the Earth moves

as expressing a relation between Galileo and a sentence type, viz. 'the Earth moves'. Instead, Davidson proposes that we view each use of such a sentence as expressing a relation (the 'samesaying' relation) between Galileo and the utterance following the word 'that'. More precisely, Davidson urges that the word 'that' be construed as a demonstrative, referring to the speech act which follows.[9] The speech act is, in the non-philosophical sense, an act, a sort of 'skit' produced not as an assertion, but as a demonstration. It is analogous to the role of an obscene gesture that might accompany my utterance of

When the President thought the TV camera was off, he went like that.

The analogy with a gesture is an illuminating one, since the type of gesture performed is very much a function of the surrounding context. Thus if I say 'The President went like that' and accompany my utterance with a sweeping motion of my open palm, the gesture I am attributing to the President will vary dramatically depending on whether the motion of my hand stops in mid air or adjacent to the cheek of a man in my audience. Similarly, in the speech act (the skit or demonstration) that follows my demonstrative, questions of ambiguity and reference will be largely resolved by the setting in which the act takes place, just as they would be were the act to have been performed in earnest.

Now let us apply this Davidsonian idea to the case of belief-attributions.[10] The first step is to give up the Carnapian picture of the logical form of belief-attributions in favour of a Davidson-style demonstrative account. So let us replace (5) with something like

(6) $(\exists b_1)(\exists b_2)$[John is in b_1 & (John, b_1, t_n) is a belief-state
    & (I, b_2, t_n) is a belief-state & B((I, b_2, t_n), that) &
    Content Identical ((John, b_1, t_n), (I, b_2, t_n))].
    Snow is white.

In (6), the word 'that' in the fourth conjunct is to be taken as a *demonstrative* referring to the performance that follows. 'Snow is white' is the script for the skit, the pretend assertion to which the demonstrative refers. In making the change, we must also adapt our construal of the B relation. As set out in the previous section, B was a relation obtaining between a belief-state and a sentence type when the belief-state could play the central role in a typical causal history of an utterance of the sentence. But if we follow Davidson's lead, the B relation we need must obtain between a belief-state and the play acting assertion of the content sentence to which the demonstrative 'that' refers. These play acting assertions serve the function of exhibiting the sort of verbal behaviour that a belief, content identical with the one being attributed to John, typically causes in the speaker. The play acting assertion itself is not caused (in the typical way) by a belief-content identical with John's. Rather, we are making a *counterfactual* claim about the content skit. If it *had* been made in earnest then it would have been caused by a certain belief-state.[11] Since it was not made in earnest, the belief-state need not even exist in the speaker. It may be only a *possible* belief-state, the possible state which would have played the central role in a typical causal history of the play acting utterance, had it been made with such a typical causal history. So the relation B will be the one which obtains between a possible belief-state and a play acting assertion if and only if the following condition obtains:

were the play acting assertion to have had a typical causal history, then the state in question would have played the central role in that history.

To finish our work we need only stipulate that the third occurrence of '(I, b_2, t_n)' in (6) also be taken as referring to a possible belief-state, or, what amounts to the same thing, that the clause in which it is embedded be construed counterfactually. In so doing we will have eliminated any remaining implication that the speaker must share the beliefs he attributes to others.

V. Content identity and content similarity

Midway through the previous section the notion of *content identity* made its appearance in my analysis. At the time I issued a promissory note to cash the concept in creditworthy coin. It is time to start paying off on that note. As a beginning, let us recall just what work the notion of content identity was required to do for us. I began with the assumption that the gross architecture of our intuitive belief–desire psychology was adequate to characterize a class of physical states that were functioning as belief-states in the individuals to which the theory applied. These are the physical states of the individual that interact with perceptual inputs, with other belief-states, with physical states functioning as desire-states in the individual, with behaviour of the individual, etc., in the ways specified by the gross architecture of the belief–desire theory. The basic theme in my account of content ascription is that when we attribute a belief to a person by specifying its content, we are identifying the belief by associating it with a possible belief-state of our own. In uttering the content sentence, we are exhibiting the characteristic effect of that possible belief-state-token. But having succeeded in picking out the appropriate possible belief-state of our own, we must still relate it to the belief-states of the person to whom we are attributing a belief. It is just here that the notion of content identity was pressed into service. Content identity was taken to be a relation among belief-state-tokens which groups them into equivalence classes. Intuitively, all the members of such an equivalence class should have the same content. Then to say that a subject believes that p can be construed as saying that among the subject's belief-states is one which is content identical with the possible belief-state indicated by our play acting utterance of p. Since our common-sense psychology plainly allows that organisms which are physically quite different may both believe that p, the physical identity of a pair of belief-state-tokens is neither necessary nor sufficient for content identity. What is needed instead is some account of how belief-state-tokens which may be physically quite different from each other are to be collected together into content identical equivalence classes.

In addition to its overt role in (6), the notion of content identity was pressed into service less conspicuously in elaborating our account of the notion of a *typical causal pattern*. In urging that there is a typical causal pattern underlying most of the assertions we and our co-linguists make, we do not expect that the pattern will be manifest at the level of physical description. Rather, we expect the pattern to emerge when we ignore the differences between content identical physical states. So, in working toward an account of content identity, we had best keep in mind this second function that the notion must serve.

Now it might be thought (and indeed I once thought) that explicating the needed notion of content identity ought to be a relatively straightforward matter. After all, we have as our data our intuitions about a broad range of cases, real or imagined, where we judge that a pair of subjects either do or do not have the same belief. To construct a notion of content identity we need only seek the principle or principles that determine these judgements. However, when we set about collecting these data, as I will in the following section, a surprising fact emerges. It turns out that there is no sharp intuitive boundary between those pairs of belief-state-tokens that we are prepared to count as content identical and those that we are not. The problem is not merely that the boundary is a bit fuzzy on the edges, which might well be expected for almost any intuitive distinction. Rather, the fuzz overwhelms the clear positive cases. The cases in which intuition dictates that a pair of belief-state-tokens clearly are content identical are swamped by those where intuition delivers an equivocal verdict. What is more, much the same phenomenon is to be found when we focus on our intuitions about the content of an individual belief-state. There too it appears that the examples in which a subject clearly believes that p are outnumbered by the cases in which intuition inclines us to say that the subject 'sort of does believe that p and sort of doesn't'. The pervasive indeterminacy of our intuitions about content and content identity is a phenomenon which any explication of content identity must capture and explain. What might the explanation be?

As a start toward answering this question, let us consider another sort of intuitive judgement where we would expect to find a few clear positive cases and a broad range of cases where intuition is indeterminate. Suppose we set a large group of subjects the task of imitating President John Kennedy's first inaugural address, having first provided them with the text. It is our task to watch the imitations, after having refreshed our memories by viewing a film of Kennedy's inaugural. After each imitation we must judge whether it is *similar* to the original. What sort of judgements would we expect to find? Well, if the subject pool included a particularly gifted mimic we would find ourselves strongly inclined to judge that his performance was similar to Kennedy's. At the other extreme, the subject pool may well include some people who are singularly bad at duplicating accents, gestures and rhythms of speech. In these cases we would find ourselves strongly inclined to say that the performance is not similar to Kennedy's. Between the extremes there will be a broad range of performances where our intuitions will be quite equivocal. The point I am labouring in this example is that everyday judgements about similarity exhibit a broad range of fuzziness or indeterminacy quite analogous to the pattern exhibited by our intuitions about the content and content identity of beliefs.

The parallel between similarity judgements and content judgements suggests that some appropriate notion of similarity between belief-state-tokens would likely play an important role in explicating the notion of content identity. And this in turn suggests that we might try to make do in our analysis merely with the required notion of similarity, and let content identity go by the board. That is, we might replace (6) with the following variation:

(7) $(\exists b_1)(\exists b_2)$[John is in b_1 & (John, b_1, t_n) is a belief-state & (I, b_2, t_n) is a belief-state & B((I, b_2, t_n), that) & Content Similar ((John, b_1, t_n), (I, b_2, t_n))].
Snow is white.

Here, *Content Similar* is the appropriate similarity relation among belief-state-tokens that remains to be explicated. The rest of the analysis is to be interpreted as in (6), with the demonstrative 'that' again referring to the play acting utterance of the content sentence whose script is provided. If, as I propose, we adopt (7) in place of (6) we will have eliminated the principal occurrence of the notion of content identity in our analysis. We can eliminate appeal to content identity entirely by replacing it, in the account of typical causal pattern, with the notion of a relatively high degree of content similarity. That will inject an element of vagueness into the account of typical causal pattern, but this will do no harm. We must now give some account of *content similarity*.

It might be objected that we seriously distort our account of content attribution by incorporating a notion of similarity. For similarity, after all, is a graded notion, a matter of degree. In general, it makes good sense to say that A is more similar to B than to C. But, the objection continues, in specifying the content of a belief we do not ordinarily take ourselves to be making a claim that admits of degrees. However, I think that this objection poses no serious threat to our emerging account of content ascription. For nothing in the account entails that content ascriptions are ordinarily taken to be a matter of degree, nor that we are aware of making the similarity judgements that underlie judgements of content. To see the point, consider the (quite intentional) analogy between the account I have been urging and *prototype theories* in recent cognitive and developmental psychology. Those theories are attempts to characterize the sort of mental mechanisms that underlie subjects' judgements about the application or non-application of a concept. The basic idea of prototype theories is that subjects have a stored paradigm or prototype, a mental representation of a prototypical instance of the concept. When required to judge whether a given object falls under the concept, subjects determine the *similarity* between the object at hand and the prototype, and it is this similarity which determines their judgement. Thus, for example, in judging whether a given object is a cup, the theory claims the subject will compare the object to his stored representation of a prototypical cup. If the object is sufficiently similar, the subject will judge affirmatively; if not, the subject will judge negatively. There will also be a substantial intermediate range where the subject finds it difficult to decide (cf. Rosch 1973a, 1973b, 1975, 1977; Rosch and Mervis 1975; Tversky 1977). Advocates of the prototype theory do not claim that subjects are aware of making similarity judgements when they classify objects as falling under a concept. The judgements are made unconsciously. Moreover, the subject need not ordinarily view the concept in question as one which admits of degrees. We commonly think that a given object either is a cup or is not, and it is a bit odd to say that one object is more of a cup than another. Yet there are many objects about which our judgements are indeterminate, and when a pair of these objects is in question, we may well be inclined to say that one of them is more cuplike than the other. Analogously, as we shall see in the following section, there are many cases where we have no clear intuition about whether or not a given subject's belief counts as a belief that p. But if we attend to a pair of such cases, we will often be inclined to say that one of the two is more nearly a belief that p than the other.

Though invoking a notion of similarity in our analysis of content ascription does not entail that ascriptions of content are ordinarily thought of as matters of degree, it does entail that content ascriptions are to some extent vague. This is a claim that is likely to provoke objections. For it is widely held that sentences of the form *S believes that p*, at least when interpreted in the *de dicto* sense, say something exceptionally sharp and precise about S. On this view, even the subtlest change in the content sentence profoundly changes the claim being made about the subject, indeed so profoundly that the original claim may be true while the altered claim is false.[12] I think, however, that those who urge the precision of content-specifying belief-attributions are being misled by a phenomenon which I shall call the *pragmatic sensitivity* of belief-attributions.

To explain what I have in mind, let me recur to the example of the attempts to imitate President Kennedy's inaugural. As I recounted the case, there was no particular purpose for which the similarity judgements were being made. And this is not all that unusual. We sometimes make or are called upon to make similarity judgements with no particular purpose in view. But we are also often called upon to make similarity judgements when there is some plan in view or some further question that is to be answered. And in these cases our similarity judgements may be importantly influenced by the pragmatic surround of goals and further questions, whether or not we are consciously aware of the influence, or able to articulate it. So, for example, in the pragmatic vacuum of the imagined experiment as I first described it, we might well judge that a gifted woman mimic's performance was very similar indeed to Kennedy's, even though her voice was several octaves higher than Kennedy's. If, on the other hand, the announced goal of the exercise was to find someone to perform Kennedy's inaugural in a radio play, then the woman mimic's performance might strike us as not particularly similar to Kennedy's. Though with this goal in mind, we might give high marks to the performance of a man whose voice and accent were suitable, though his gestures were all wrong. However, his performance would not rate as similar either, if the goal in view were to find a stand-in for Kennedy who would give a televised speech while the real President Kennedy was at a secret meeting.

Now it is my contention that the similarity judgements which underlie our intuitions about content are analogously affected by the pragmatic surround in which the content ascription is called for. Our judgements about the similarity of belief-states, and thus about the contents of beliefs, are significantly affected by the explanatory or predictive use to which our belief-attribution will be put. Of course, the analogy between similarity judgements in the imitating Kennedy case and the similarity judgements which I claim underlie content ascriptions must be taken with a grain of salt. There are significant differences between the two cases.

Principal among them is that in the imitation example we are consciously aware that we are making a similarity judgement, while in the content judgement case the similarity judgement is sub-doxastic; we are unaware of making it. Also, in the imitation case and in other cases of explicit similarity judgements, we are often at least partially aware that the items being judged are similar in some *respects* and not in others. Thus we can, and often do, have some awareness of the workings of pragmatic factors on our judgements. We are aware of emphasizing those features or respects that are relevant for the matter at hand, and de-emphasizing others. In the case of content judgements, we generally have little or no awareness of the influence of the pragmatic surround. Though, as I shall show in the following section, we can demonstrate the influence of pragmatic factors by varying them and noting how our intuitions about content change.[13]

Finally, let me explain how I think pragmatic sensitivity of content judgements can create the illusion that content-specifying belief-attributions say something very precise about the believer. To create the illusion we first restrict ourselves to cases where the belief-states in question are those of our compatriots, people whose beliefs are very similar to our own in all those respects or components of similarity to be adumbrated in the following section. We then focus our pragmatic concern on predicting or anticipating the *verbal* behaviour of our compatriot: what, precisely, will (or would) he say? When we pose a question about the content of a belief in this setting, small differences will loom large. Any difference between belief-states that will reflect itself in the words the subject will choose to express his belief will count as difference enough to make the beliefs dissimilar. But when the believer is more exotic, a person from quite a different culture perhaps, or a child, and we are less concerned to anticipate exactly what he will say, then small differences pale. The imitating Kennedy example once again provides a rough analogy. If the project at hand is locating a double to stand in for Kennedy, deceiving the press and the public, then small differences will loom large. In this frame of mind we might be tempted to think that in saying a given subject's performance is similar to Kennedy's we are saying something very precise. We might even, perhaps, be tempted to say that positive similarity judgements in other contexts invoked different senses or 'similar' or perhaps used 'similar' in a metaphorical extension of its basic sense. Analogous claims have been made about the ambiguity or metaphorical use of 'believes' when applied to animals, children or exotic folk. By my lights they are no more plausible.

VI. Toward a theory of belief-state similarity

In this section I will pursue a pair of interconnected projects. First, I will begin work on a theory about the notion of belief-state similarity—the notion which, I have been urging, we invoke when we make judgements about the contents of beliefs. As a step toward an account of belief-state similarity, I will characterize four distinct respects in which belief-states may be similar to or different from each other, which are relevant to our judgements about content. Much of what I have to say in this section will be aimed at explaining these four dimensions of similarity and illustrating some of the ways in which they influence content judgements. But my remarks will be no more than the tentative beginnings of a theory. I am not at all sure that the four dimensions of similarity I will consider are the only dimensions that influence our content judgements.[14] Moreover, what I have to say about the four will still leave much unsaid. Three of the four components of similarity on my list are notions which, quite overtly, admit of degrees. And I suspect the fourth would also turn out to conceal a graded notion if analysed in more detail that I shall attempt here. So a complete account of each component would have to include some explicit story about how the magnitude of each is to be computed. And for an account of the notion of *overall* belief-state similarity (the notion which actually functions in my analysis) we would also require an explicit account of how the four components are weighted to produce an overall judgement of similarity. This latter story will be a particularly complex one since the weighting given to the several dimensions will not be constant. Rather, as I urged in the previous section, the judge's assessment of the pragmatic features surrounding his judgement will determine which components of overall similarity are important and which are not. I will make no attempt to provide an explicit account of how the magnitudes of the several components of belief-state similarity are to be computed, nor will I offer a worked out theory of how these various magnitudes yield an overall judgement of similarity.

The second project in this section is to illustrate what I earlier called the *pragmatic sensitivity* of belief-attributions. What I will try to show is that if we alter the importance of a component of similarity in the situation surrounding a belief-attribution, we can alter our intuitions about the acceptability of the content ascription. If the interests of the speaker and/or audience make a high degree of similarity along a particular dimension a matter of little moment, then we will find certain belief-attributions to be perfectly acceptable. But if the interests of the speaker and/or audience make a high degree of similarity along that dimension important, then the same belief-attributions about the same people will be intuitively unacceptable.

1. Functional similarity

The first entry on my list of components of belief-state similarity is in the spirit of the much discussed notion of functional identity. I will call it *functional similarity*. The notion of a *functional theory* provides a convenient starting point in explaining the ideas of functional identity and functional similarity.

A functional theory is a theory which attempts to explain the behaviour of an organism or other complex system by postulating a number of internal states that the system may from time to time be in, and specifying the causal relations among the postulated states, environmental stimuli and the overt behaviour of the organism or system. What is characteristic of a functional theory is that it makes no commitment about the physical or physiological nature of the postulated internal states; indeed, strictly speaking, it need not even be assumed that they are physical or physiological states. Only the causal relations among stimuli, postulated states and behaviour is specified. Thus we might think of a functional theory as making an existential claim about a certain organism, viz. the claim that *there are* a number of physical or physiological properties which, when instantiated by the organism, yield state tokens whose causal interactions are as specified by the theory. A functional theory will thus have the logical form of a complex open sentence which is true of an organism just in case there are physical or physiological properties such that when the organism instantiates them, the resultant state tokens have the causal interactions detailed by the theory.

An illustration may serve to make all this a bit clearer. Suppose we take

$$(x, P_1, t) \Rightarrow (x, P_2, t + \Delta)$$

to mean

x's instantiating P_1 at t would cause x's instantiating P_2 at $t + \Delta$.

Then a particularly simple functional theory might look something like this:

$$(\exists P_1)(\exists P_2) \ \ldots \ (\exists P_n)(t)[(x, \ S_1, \ t) \Rightarrow (x, \ P_1, \ t + \Delta) \ \ldots$$
$$\& \ (x, P_2, t) \Rightarrow (x, P_n, t + \Delta) \ldots \& \ (x, P_n, t) \Rightarrow (x, B_1, t + \Delta)]$$

where 'S_1' and 'B_1' denote specific environmental ('stimulus') and behavioural properties respectively, and 'x' is a free variable.[15] More complex functional theories will commonly have conjunctive and/or disjunctive states flanking the arrow. When the details of a functional theory are of no concern, we can conveniently represent their general form as follows:

$$(\exists P_1)(\exists P_2) \ldots (\exists P_n) \ T \ (P_1, P_2, \ldots, P_n, x).$$

To say that a given organism, O, satisfies a functional theory,

$$(\exists P_1)(\exists P_2) \ldots (\exists P_n) \ T(P_1, P_2, \ldots, P_n, x)$$

is to say that there is a sequence of properties, $Pr_1, Pr_2, \ldots Pr_n$, such that the sequence

$$Pr_1, Pr_2, \ldots, Pr_n, O$$

satisfies

$$T(P_1, P_2, \ldots, P_n, x).$$

However, the sequence of properties in virtue of which one

organism satisfies a given functional theory need not be the same sequence in virtue of which another organism satisfies that same theory. Thus it may be the case that organisms which are physically quite different may nonetheless both satisfy a given functional theory, though they will do so in virtue of different sequences of properties.

A number of writers have attempted to exploit the notion of a functional theory to give a general account of the conditions that must obtain when a pair of organisms have a psychological state in common. This 'functionalist' account requires that two things be true of a pair of organisms if they are to have a psychological state in common. First, there must be a functional theory which is satisfied by both organisms. If this is true, then there will be a sequence of properties in virtue of which the first organism satisfies the theory; call them Pr_1^1, Pr_2^1, ... Pr_n^1. And there will be another sequence of properties in virtue of which the second organism satisfies the theory; call this second sequence Pr_1^2, Pr_2^2, ... Pr_n^2. Now the second condition that must be met, according to the functionalist account, if the organisms are to have a psychological state in common, is that the organisms must *currently be instantiating functionally correlated properties.* That is, there must be some i such that the first organism is instantiating Pr_i^1 and the second is instantiating Pr_i^2.

This functionalist account of psychological state identity has a pair of attractions. First, it nicely accommodates the intuition that physically quite different organisms (and even artefacts like automata) could be in the same psychological state. For, as we have lately noted, physically quite different systems may both satisfy the same functional theory. Also they might, at a given moment, each instantiate functionally correlated properties. Second, the functionalist account does justice to the idea that the identity of a psychological state is determined, at least in part, by the role it plays in a system or network of states and processes. For an organism to have a psychological state in common with another organism, it must instantiate a property which is a member of a *sequence* of properties, and the sequence in turn must satisfy the functional theory.

It is important to stress that, on the functionalist account, the mere fact that a pair of organisms both satisfy the same functional theory tells us nothing about whether they have any psychological states in common. Indeed, a pair of organisms might both satisfy the same functional theory while having *no* psychological states in common. Also, for the functionalist, the fact that a pair of organisms have one psychological state in common entails nothing about whether they have any further psychological states in common, though it does entail that they both satisfy the same functional theory. These facts will prove to be of some moment in the subsection that follows.

As a final preliminary, let me introduce the notion of *functional identity*. Functional identity is a relation between a pair of physical state tokens. Intuitively, a pair of physical state tokens are functionally identical if they are instantiations of functionally correlated properties, in the sense of this term introduced three paragraphs back. More formally, the states (S_1, P_1, t) and (S_2, P_2, t) are functionally identical if there are two sequences of properties, $Pr_1^1, Pr_2^1, \ldots, Pr_n^1$, and $Pr_1^2, Pr_2^2, \ldots, Pr_n^2$, such that the former sequence along with S_1 satisfies a given functional theory, the latter sequence along with S_2 satisfies the same functional theory, and for some i, P_1 is identical with Pr_i^1 and P_2 is identical with Pr_i^2. Technically, functional identity as here defined is a *three* place relation: a pair of states are functionally identical with respect to a functional theory. This technicality will return to trouble us shortly.

Now recall that our current project is to characterize the relation of belief-state similarity. It is my contention that one of the dimensions along which the overall similarity of belief-states is assessed is *functional similarity*, where functional similarity is construed as approximation to functional identity. So, subject to sundry quibbles soon to be raised, a pair of functionally identical belief-states count as maximally functionally similar. This fixes the zero point on the functional dissimilarity scale. But obviously a great deal more work would be needed to say how the similarity of a pair of non-functionally identical states is to be assessed. And, as I warned earlier, this is a project which I shall decline to tackle. When it comes to marshalling examples to illustrate the sensitivity of content judgements to functional similarity, our informal intuitive assessment of the degree of functional similarity between the states described will have to suffice in place of a more detailed theory.

Let us look at some of the evidence for taking functional similarity to be one of the dimensions along which belief-state similarity is assessed. If the theory I have been urging correctly captures the general structure of the cognitive mechanism underlying our intuitions about content, then we can make a number of predictions about these intuitions. First, we would expect that by varying the manifest functional similarity between a subject's belief-state and the appropriate belief-state of our own, we should be able to vary our intuitive willingness to say that the subject believes that p. Other things being equal, the greater the functional *dis*similarity between the subject's belief-state and the belief-state which is B-related to our play acting utterance of p, the less willing we should be to count the subject's belief-state as the belief that p. Second, we would expect that by varying the pragmatic context in which a judgement is called for we should be able to vary our intuitions about the belief-state of a single subject. I think that both of these predictions fit nicely with the facts.

As an illustration of the first prediction, consider the following imaginary experiment. Suppose we have three subjects all of whom are generally fluent in English and are quite normal, apart from the anomalies to be noted. We present our subjects with a range of colour patches under standard illumination, then ask them to say what colour each patch is. We then ask them to tell us, from memory, the colour of some familiar objects, such as English mail boxes and 'Beefeaters'' jackets. Our first subject turns out to describe the colour patches just about as I would, though some of the patches that I would call 'reddish-orange' he calls 'orangish-red'. In response to our second round of questions, he says, 'Beefeaters' jackets are red'. Now the question I want to raise is whether it is intuitively natural to say that this first subject believes that Beefeaters' jackets are red. The answer, I think, is clearly affirmative.

Our second subject's colour judgements are rather more anomalous. There are a number of shades of blue and green which this subject labels 'red'. Moreover, the anomaly is not a superficial linguistic one, since in colour matching tests he regularly confuses these shades of blue and green with certain shades of red. If we show him a red colour patch of an appropriately chosen shade and ask him to reidentify it when given a choice between the original and a blue patch of a certain shade, his performance is no better than chance. However, the anomaly we are imagining pertains only to a few selected shades of green and blue. For other shades his performance is indistinguishable from that of a normal subject. Finally, let us assume that our subject is not aware of his condition. (A number of people have sceptically suggested that this last assumption is absurd, since an otherwise normal adult could not fail to know that he was perceptually anomalous in this way. However, this scepticism generally declines when the sceptic learns that colour blindness was not discovered until the late eighteenth century (Gregory 1966, p. 126).) When we ask our second subject about the colour of Beef-eaters' jackets, he replies, 'They are red'. Is it intuitively plausible to say that our second subject believes that Beefeaters' jackets are red? Here, I think, most people would still be inclined to answer affirmatively, though rather more hesitantly than in the first case. In view of the functional dissimilarity between the second subject and ourselves, this is just the result my theory would lead us to expect.

For a final subject, imagine a person whose colour perception is still more anomalous. This subject, like the previous one, labels certain green and blue patches 'red'. But he also calls some red patches 'blue'. As in the previous case, he cannot reliably reidentify an appropriately selected red patch mixed with a selection of blue ones. But, in addition, this subject calls certain objects 'red' regardless of their colour, provided that they have certain *shapes*! Thus he will label 'red' any object which is roughly the shape of a map of Britain and also any object which is roughly the shape of a Volkswagen Beetle. When taken to an auto dealer's lot full of VW Beetles, this subject reports, 'They all are red'. If one of these cars is singled out (say a bright yellow one), then mixed in with the others, he cannot reidentify it. He says, 'They all look the same'. Finally, let us assume that, as in the previous case, the subject is unaware that his colour judgements are anomalous. When we ask this subject about the colour of Beefeaters' jackets, he too says, 'They are red'. Does he believe that Beefeaters' jackets are red? Here, I find, most people report that they have no clear intuitions. We are inclined to say that he 'sort of does believe that they are red', but then we are also inclined to say that he 'sort of doesn't'. On the theory I have been elaborating, these are just the intuitions we would expect.

Let me turn now to the pragmatic sensitivity of our content intuitions. The third, and most anomalous, of our imagined subjects can be used to construct the sort of example we need. Suppose that this subject, Mr Oddsee, to give him a name, has been witness to a murder which has all the earmarks of a Mob hit. On being questioned by the police, Oddsee describes the culprit and reports that he fled in a red Volkswagen Beetle whose licence plate ended in the letter 'X'. Lt Smith who is in charge of the police investigation, alerts his men to investigate all red VW Beetles whose licences end in 'X'. However, shortly after this, a psychologist who is aware of Oddsee's anomalous perception reveals this information to Smith, who promptly cancels the order to investigate red VW Beetles. His superior, unaware of Oddsee's anomaly, asks why the hunt for the red Beetle has been called off. 'Didn't the witness say that the killer escaped in a red Beetle?' he asks. Here we can well imagine Smith replying, 'Yes, but I don't really know whether he believed it or not.' Smith's reluctance to say that Oddsee believes the killer escaped in a red VW Beetle is natural enough under the circumstances. Whether or not a witness believes that the killer escaped in a red car is of interest to Smith because if a witness does believe it this is some evidence that the killer in fact did escape in a red car. But because of Oddsee's functional dissimilarity with most of us, his belief is of no use in determining the colour of the get-away car. With more normal subjects, there is a generally reliable causal connection between seeing a red car drive away and coming to believe that a red car drove away. But in Oddsee's case the causal link between stimulus and belief does not obtain. Since it is this specific functional feature of the normal belief which is important in our story, we, along with Smith, are strongly inclined to deny that Oddsee believes the killer escaped in a red VW Beetle.

If, however, we pose the question with different purposes in mind, we can succeed in provoking the opposite intuition. Suppose that there was another witness to the crime, one whose colour vision is quite normal. He believes correctly that the get-away car was a bright yellow VW Beetle. This second witness, call him Mr Coward, is afraid to go to the police and tell them what he knows. He fears Mob retribution. When he reads in the newspaper that the other witnesses said the get-away car was red, Coward is puzzled and upset. Perhaps Oddsee wants to deceive the police, he thinks. Perhaps he is even tied in with the Mob. But Mr Coward's qualms are soon quieted. He confides in a psychologist friend, the very one who told Smith about Oddsee's functional anomaly. On hearing the psychologist's account of Oddsee's performance in colour labelling tasks, Mr Coward says, 'Oh, now I understand why he said what he did. He actually believed that the get-away car was red.' In this context, I think Coward's claim about the content of Oddsee's belief is intuitively natural. What interests Mr Coward is why Oddsee said what he did. The belief which plays a central role in the explanation interacts with other beliefs (e.g. beliefs about what Smith is asking), with desires (e.g. the desires to tell the police what happened) and with behaviour in much the same way as the belief we would express by saying 'the get-away car was red'. Since these are the similarities that are important in this context, we find it intuitively acceptable to describe Oddsee's belief as Coward does.

So far I have been concentrating on examples of subjects whose functional peculiarities are perceptual. However, a parallel array of intuitions would emerge if we were to attend, instead, to examples of subjects whose *inferential* patterns differ from our own. If a person is a bit slower than we are at noting the logical implications of his beliefs, or a bit more inclined to draw invalid inferences, but is otherwise suitably similar to ourselves, we will have little trouble ascribing content to his beliefs. But as the pattern of inferences he exhibits becomes increasingly dissimilar to ours, our intuitions about the content of his beliefs will grow increasingly weaker. In the extreme case we will find ourselves quite incapable of telling any intuitively plausible story about the content of the subject's beliefs.

2. *Ideological similarity*

The second entry on my list of features that determine belief-state similarity is what I shall call *ideological similarity*. Perhaps the best way of explaining the notion is to contrast it with functional similarity. Recall, the fact that a pair of states are functionally identical entails nothing about whether the organisms in these states are in any further functionally identical states. The parallel claim holds for functional similarity. Thus it is possible for a pair of belief-state-tokens to be functionally similar or functionally identical even though the organisms in these states are in no other states which are functionally similar or functionally identical. However, such a pair of belief-states would not be judged to be similar in content. For in assessing the content similarity of a pair of belief-state-tokens we consider not only the functional similarity of the states themselves, but also the functional similarity or dissimilarity of the *other* belief-state-tokens that the organisms happen to be in. As the surrounding beliefs become increasingly less similar, a pair of *functionally* similar belief state-tokens will count as increasingly less similar *in content*. Thus the content we ascribe to a given belief-state-token will be determined in part by the other beliefs the organism is in.

This can be said a bit more carefully. First consider the extreme case where each current belief-state-token of one organism is functionally identical with a current belief-state-token of a second organism, and vice versa. I will say that such organisms are *ideologically identical*. A weakening of this notion yields a relation of ideological similarity. A pair of organisms will count as *ideologically similar* if a substantial number of the current belief-state-tokens of one organism are functionally similar to current belief-state-tokens in the other organism, and vice versa. Ideological identity, then, fixes the zero point on the scale of ideological dissimilarity. However, this notion of ideological similarity is a relation between organisms, while for our purposes we require an ideological similarity relation between belief-state-tokens. As a first pass at defining that notion, we might try the following: A pair of belief-state-tokens (S_1, P_1, t_1) and (S_2, P_2, t_2) are ideologically similar iff (or to the extent that)

   (i) (S_1, P_1, t_1) and (S_2, P_2, t_2) are functionally similar, and
 (ii) S_1 at t_1 is ideologically similar to S_2 at t_2.

Put less formally, a pair of belief-state-tokens are ideologically similar if they are functionally similar and are states of ideologically similar organisms.

Now it is my contention that a notion of ideological similarity along the lines of this one is invoked in assessing content similarity, and thus in ascribing content to belief-states. However, even if we forgive its vagueness on various matters, it is pretty clear that this notion of ideological similarity between belief-states is not quite the one we actually invoke. As I have explained it, the ideological similarity of a pair of belief-state-tokens is a function of the ideological similarity of the subjects who are in the belief-state-tokens. But our actual assessments of ideological similarity are more subtle. Not all parts of your store of beliefs are equally relevant in judging whether one of your beliefs which is functionally similar to one of mine is also ideologically similar. Thus it may be that you and I both believe that wood floats on water, even though you share few of my beliefs about free will, frisbee or the history of France. Indeed, you might have no beliefs at all on these last three matters, and it would not incline me to doubt that you believe wood floats on water. But if you share few of my beliefs about wood, floating and water, then I will be little inclined to say that you believe wood floats on water. And if it is urged that you believe wood floats on water though you have *no* other beliefs about wood, floating and water, intuition would find the suggestion quite incomprehensible. It appears, then, that in assessing the ideological similarity of belief-states, other beliefs built from the same concepts are generally weighted more heavily.

If I am right about the role of ideological similarity in the processes underlying content ascription, then two predictions follow concerning our intuitions about content. First, we should expect that when judging if a belief-state of a given subject is the belief that p, our views about what additional, related beliefs the subject has should play an important role. The greater the difference between his store of related beliefs and ours, the less inclined we will be to count his belief-state as the belief that p. Second, we should expect that the degree of ideological dissimilarity which is compatible with an intuition that a belief-state is the belief that p will be determined in part by the explanatory and communicative circumstances (the 'pragmatic surround') in which the question arises. Both of these predictions accord with the facts.

As a first example, consider the shrinking cognitive world of the person afflicted with degenerative senility. The condition is one marked by progressive and generally irreversible loss of memory. In advance stages of the disease, a sufferer will sometimes retain the capacity to report an isolated fact while almost all the surrounding network of beliefs has been lost. In one case known to me, the subject had, as a young woman, been much distressed by the assassination of President William McKinley. In middle age she often recounted her memories of the days following the assassination, and readily elaborated her analysis of the impact the assassination had had on the politics of the day. In old age, when her degenerative senility had become quite severe, she would still respond to the question, 'What happened to President McKinley?' by saying, 'He was assassinated', much as she would have before the onset of her disease. But when queried further about the event, she could give few answers. Asked who McKinley had been, or what an assassination was, she shrugged and apparently could give no answer. Sometimes she would reply, 'I don't remember'. Asked whether McKinley was dead, she said she didn't know. No doubt the affliction from which she suffered had taken its toll on the functional organization of her cognitive system. But the most noticeable symptom of her disease was the massive loss of memory or belief. Now it is beyond question that when the subject was younger she had believed that McKinley was assassinated. Did she believe it when the degenerative senility had become quite advanced? I think that most people will be strongly inclined to say no; by the time the disease had progressed to the point where she no longer knew that assassinated people die, she could no longer comfortably be said to believe that McKinley was assassinated. However, there is no point in the progress of the disease that marks a sharp boundary between the time she did and the time she did not believe that McKinley was assassinated. Instead, what we find is a gradual change in our intuitions. The more her memory deteriorates, the more related beliefs she loses, the less we are inclined to say that she believes McKinley was assassinated. This is just what we would expect on the view I have been urging about the role of ideological similarity in content ascription.

For a second example, consider the subject of A. R. Luria's gripping case history, *The Man With a Shattered World* (1972). Luria's patient, Zasetsky, suffered a severe head wound while fighting with the Red Army during World War II. Though the injury took its toll on a broad spectrum of his cognitive functions, Zasetsky was most troubled by his massive loss of memory and knowledge. 'I remember nothing, absolutely nothing!' he wrote. 'Just separate bits of information that I sense have to do with one field or another. But that's all! I have no real knowledge of any subject. My past has just been wiped out!' (Ibid., p. 116.) Luria describes this effect of the wound as a partial destruction of Zasetsky's conceptual networks.

> [Zasetsky] referred to his major disability as a loss of 'speech memory'. And he had good reason to do so. Before he was wounded, his words had distinct meanings which readily occurred to him. Each word was part of a vital world to which it was linked by thousands of associations; each aroused a flood of vivid and graphic recollections. To be in command of a word meant he was able to evoke almost any impression of the past, to understand the relationships between things, conceive of ideas and be in command of his life. And now all of this had been obliterated. [Ibid., p. 89.]

With difficulty, Zasetsky could verbally identify some of the objects, persons and substances he had known before the war. But his massive loss of memory makes it unclear just what to say about the beliefs he was expressing when he made these verbal identifications. Thus he could apparently report that certain common objects shown to him were made of wood. Yet he wrote, 'I can't understand how wood is manufactured, what it is made of. Everything—no matter what I touch—has become mysterious and unknown.' (Ibid., p. 87.) Does a man who can often correctly identify wood when he sees it, but who has lost most of his knowledge about wood, really believe that the objects he identifies are wood? Intuition, surely, leaves the question moot. And this is just what would be expected by the theory I am defending.

By a fanciful elaboration on Luria's account of Zasetsky we can illustrate the predicted pragmatic sensitivity of our intuitions about the beliefs of people who are ideologically exotic. Let us imagine that as part of this therapy Zasetsky is taught a game. The object of the game is to find some trinket hidden in the room. To help his search he is given various types of objects which have previously been assigned conventional significance as clues. Water, for example, means that the hidden trinket is near the sink; paper means that it is in the closet; wood means that it is hidden somewhere up high, etc. Now suppose that in the course of playing this game with Zasetsky we give him as a clue a piece of Formica with a wood-grained pattern. Zasetsky immediately heads for the ladder, climbs it, and begins hunting around on the tops of the bookcases. Why? Here I think it would be perfectly natural to say that Zasetsky believed the clue was made of wood, and he inferred that the hidden trinket was somewhere up high. The plausibility of attributing to him a belief with that content is little affected by our knowledge of the ideological gap between Zasetsky and ourselves, since in this case the ideological difference makes little difference for the explanation being offered.

As a contrast, let us imagine a rather different game. In this second game we show Zasetsky two objects; one is a piece of lumber, the other is a cloud of CO_2 gas rising from a tub of water containing a chunk of dry ice. We ask Zasetsky which of the objects is wood, and he points to the piece of lumber. We then set a second question for him. We tell him that one of the two objects before him is made from the trunk of a tree, is solid and quite strong, is often used for building houses and furniture, can be shaped with a sharp knife or axe. The other object, we tell him, is not solid at all; he could pass his hand through it easily. It can be inhaled or blown away with a puff of air. Having told him all this, we ask him which is which. Is it the wood which is made from the trunk of a tree, etc? Or rather is the wood the one which can be inhaled, blown away, etc? His answer is that he does not know. In the wake of his inability to answer our question, are we prepared to say he believes that the lumber he initially pointed toward is wood? Here, I think, intuition inclines us much more strongly toward a negative answer.

Both Zasetsky's case and the degenerative senility case involve subjects whose store of beliefs is different from our own largely in virtue of being *smaller*. In each case the subject started out with a collection of beliefs reasonably similar to ours, and, as the result of illness or injury, *lost* many of his beliefs. However, I think the situation is quite parallel when we attend to subjects whose store of beliefs is, from our perspective, increasingly *exotic*. As the gap between our beliefs and the subject's grows greater, it becomes increasingly difficult to say just what it is that the subject believes. Thus the historian of science or the student of alien cultures is often impelled to tell a complicated story about the practices and traditions of his subjects, and to interweave it with an account of the network of their beliefs and the words they themselves use to express these beliefs. The function of the background information is to illuminate the sort of possible world in which we ourselves might utter the native's content sentence in earnest. Only then can we make sense of the counterfactual which starts, 'If I were to have uttered that (the content sentence) in earnest, then . . .' And, on my account just such a counter-factual is embedded in the analysis of content ascriptions.

The role of ideological similarity in content ascription sheds considerable light on the much debated issue of the possibility of alternative conceptual schemes. These who deny the possibility of radically different conceptual schemes argue that no evidence we could possibly have about a creature would count as evidence that it had a radically different conceptual scheme. For if the creature's putative conceptual scheme is really *radically* different from our own, then presumably we do not have the conceptual and linguistic resources to say what the creature believes; we can ascribe no content to the creature's putative beliefs, nor translate its putative utterances. But then what reason have we to think that the creature has beliefs at all, or that its utterances have any meaning? A creature with a radically different conceptual scheme, the argument concludes, is in principle empirically indistinguishable from a creature with no conceptual scheme— no system of beliefs—at all (Stroud 1968; Davidson 1974a). Perhaps the most telling rejoinder to this argument is the observation that we ourselves, or our descendants, might hold a radically different conceptual scheme through a process of gradual evolution. By slowly replacing one belief or theory by another we can imagine the gradual emergence of a system of beliefs having little or nothing in common with the system we now hold (Rorty 1972).

Some writers have been inclined to see a genuine antinomy lurking in the head-on clash between these two aruments. However, I think the story I have been telling about content ascription paves the way for a resolution which gives each side in the debate much of what they want. Since ideological similarity plays an important role in content ascription, the opponents of alternative conceptual schemes are surely right that if a system of belief-states were radically different from our own, we would generally find ourselves quite unable to ascribe any content to them, and thus unable to say *what* conceptually exotic subjects believe. Actually, an important caveat must be added here. Ideological similarity is only one of the dimensions contributing to an overall assessment of content similarity. A pair of belief-states which are quite radically ideologically dissimilar may be much more similar along other dimensions contributing to overall content similarity. What is more, which dimensions count most heavily in an assessment of content similarity (and thus in the ascription of content) will be a function of the 'pragmatic surround'. So it will often be possible to ascribe content quite naturally to the belief-states of creatures who are ideologically quite distant from ourselves. Thus it is that we can often ascribe content to the belief-states of animals without engendering any feeling of intuitive oddness. The very same content ascription may strike us as strained and counter-intuitive, however, when the pragmatic circumstances focus attention on the ideological disparity between ourselves and the beast in question.[16]

On the other side, those who urge the possibility of alternative conceptual schemes are surely right that a creature or system could retain the global architecture prescribed by our intuitive folk psychology though its actual store of belief-states is radically different from our own. Moreover, we might well construct *and empirically* test a functional theory of the behaviour of the system—a functional theory which exhibited the gross architecture of our folk psychology and thus attributed to our hypothetical subject a sizeable store of belief-states. Though of course we would generally find it unhelpful or intuitively unnatural to ascribe content to the belief-states the theory postulates.

At this point the obdurate opponent of alternative conceptual schemes might protest that these ideologically exotic systems of 'belief-states' constitute no real challenge to his view. A belief, he might insist, must be a belief *that something or other*.[17] So, no matter how much these ideologically exotic systems of states may resemble beliefs in their gross functional properties, they simply do not count as *beliefs*. I am inclined to think that our intuitive notion of belief simply will not support this move, however. As I see it, belief-states to which no content can comfortably be ascribed are anomalies largely unanticipated by our pre-theoretic concept of belief. Our concept thus provides us with no clear decision on whether these anomalies are just peculiar beliefs, or are so peculiar as to be no beliefs at all. Were we pressed to make a decision, my hunch is that the apparent *adaptiveness* of the creature in question would loom as a major consideration. If it reproduced its kind, resisted our aggression, thwarted our attempts to capture it, and seemed to thrive after its fashion, I suspect we would count its belief-states as real beliefs despite our inability to assign them any content. Conversely, if its behaviour seemed erratic, pointless and utterly maladaptive we would be more inclined to deny it had genuine beliefs. And to the extent that it was simply unclear whether or not the creature's behaviour was adaptive, the question of whether or not it really had beliefs would be moot.

3. *Causal convergence*

Both functional similarity and ideological similarity are compatible with the spirit of a view that has of late been much discussed under the label of *methodological solipsism*. What the methodological solipsist urges, put very roughly, is that psychological states and relations, or at least those that should be of concern in a serious scientific psychology, must supervene on the current, internal, physical states of the organism(s) in question (cf. Fodor 1980; Stich 1978b, 1980). Once the current, internal, physical states have been fixed, the psychological states and relations have been fixed as well. The history of an organism will of course be relevant in explaining how the organism got into its current, internal, physical states, and thus to explaining how it got into its current psychological state. But this causal history, according to the methodological solipsist, is not relevant to a specification of *what* psychological state the organism is in; any alternative history that left the organism in the same current, internal, physical state would leave it in the same psychological state as well. Similarly, the social setting in which an organism may happen to be is not relevant to a specification of its current psychological states. An organism in a quite different setting, but in the same current, internal, physical states would be in the same psychological states as well. I think there is much to be said for cleaving to the strategy of methodological solipsism in empirical psychology, though I will not try to argue the point here. What is important for our current concerns is that the two items remaining on my list of components that contribute to our judgements of content similarity among belief-states, and thus to our judgements about content, violate the precept of methodological solipsism. In assessing the content similarity of a pair of belief-states we take account of both the causal histories of the belief-states in question and the social settings of the subjects. I will take up the point about causal histories in this subsection; social (and particularly linguistic) settings will be the theme in the subsection that follows.

Perhaps the clearest way to see how the causal history of a belief-state can influence our judgement about its content is to consider a variation on an example proposed by Hilary Putnam. Suppose that in some far off corner of the universe there is a planet, call it 'Yon', strikingly similar to our own Earth. Though evolving quite independently of events on Earth, the people of Yon have developed political and social institutions that are all but indistinguishable from our own. Even their languages are similar to Earth languages, so similar, in fact, that an American Earthling visiting that part of Yon which Yonder folk call 'America' would have no trouble communicating with the locals. Let us fancy the similarities between Yon and Earth to be so great that there are a pair of people, one on each planet, who are, at a given time, molecule for molecule replicas of each other. Both of them go by the name 'Bill Jones', and since they are physical replicas of each other, they will be functionally and ideologically identical as well. Now suppose that on Earth there is a certain religious cult leader of dubious reputation, known to the public as 'The Revd Sam'. It has been widely alleged on Earth that the Revd Sam has given large bribes to reactionary politicians, and Earthling Jones thinks these allegations are true. On Yon there is another cult leader, of equally dubious reputation, also known publically as 'The Revd Sam'. There are similar allegations about Yonder Revd Sam, and Yonder Bill Jones takes them to be true. Finally, for dramatic interest, let us suppose that the allegations of bribery are true about Yonder Revd Sam, but that Earthling Revd Sam stands falsely accused.

Now let us imagine that at a certain time we find both Earth Jones and Yonder Jones sincerely saying, 'The Revd Sam bribes politicians'. Are they expressing the same belief? I think there is a strong intuitive pull in the direction of saying that they are not. Pressed to explain our intuition, it would be natural to say that Earth Jones believes Earth Sam bribes politicians, while Yonder Jones believes Yonder Sam bribes politicians. But it would be radically counter-intuitive to say Earth Jones believes that Yonder Sam bribes anyone or to say Yonder Jones believes that Earth Sam does.

The difference in our intuitions about Earth Jones and Yonder

Jones cannot be traced to functional or ideological differences between them since, *ex hypothesi* they are functionally and ideologically as similar as can be. To explain the difference we must look instead at the *causal histories* linking a component (or concept) in Earth Jones's belief to Earth Sam, and linking a functionally and ideologically identical component in Yonder Jones's belief to Yonder Sam. It is in virtue of these quite different causal histories that Earth Jones's belief is *about* Earth Sam, while Yonder Jones's belief is about Yonder Sam. It is also (partly) in virtue of these causal histories that when Earth Jones says 'The Revd Sam gives bribes', he is referring to Earth Sam, while when Yonder Jones utters the same sequence of words he is referring to Yonder Sam. So it would appear that in assessing the content similarity of a pair of belief-state-tokens we take account of the causal histories of (at least some of) their constituent concepts. Where the concepts in question are individual concepts, the beliefs count as more content similar if the causal histories of the individual concepts *converge* on the same individual. Putnam has offered a parallel argument which suggests (though less compellingly) that the causal convergence condition applies to natural kind concepts as well (Putnam 1975a; Zemach 1976).

There is much that is unclear about the causal convergence condition on content similarity. Perhaps the most pressing puzzle concerns the nature of the required causal links between beliefs (or their components) and individuals or substances. What is obvious is that not just *any* causal link will do. To see this, suppose all of Earth Jones's beliefs about Earthling Revd Sam derive from a series of newspaper stories written by a certain journalist, and that the very same journalist, after an inter-galactic trip, launched an investigation of Yonder Sam leading to another series of articles, which were the source of Yonder Jones's beliefs. The fact that the causal histories of both sets of beliefs converge on the journalist is quite irrelevant in assessing their content. So what sorts of causal convergence are relevant in assessing content similarity? An analogous question arises for the causal theory of reference championed by Putnam and Kripke. And, unfortunately, I have little to add to the inadequate accounts in the literature.

Since the causal convergence component of belief-state content similarity is quite independent of the functional and ideological similarity components, it will sometimes happen that the various components pull in opposite directions. Thus, for example, there are many cases where causal convergence is pretty clearly not satisfied, though there is a high degree of functional and ideological similarity. In such cases my theory would predict unsettled intuitions which may be urged in one direction or another by the 'pragmatic surround'. I think there is abundant evidence that the facts are as my theory would expect them to be. However, I will forbear giving examples here. They inevitably involve the putative distinction between *de re* and *de dicto* belief-attributions, and that is a topic which demands a paper of its own.

4. *Social setting*

The final item on my list of factors that contribute to assessments of overall content similarity is what might be called *similarity of social setting*. In an important recent paper, Tyler Burge (1979a) has argued that our content ascriptions are influenced by the social and linguistic practices prevailing in a subject's community, even though crucial aspects of these practices have had no direct impact on the subject. To illustrate the point, let us consider a variation on one of Burge's examples. Suppose that on our lately imagined planet of Yon is a *Doppelgänger* of a certain Earthling, Max. Max and his Yon counterpart (also named 'Max', of course) are physical replicas of each other; what is more, they have led entirely parallel lives. For each physical stimulus that has impinged on Earth Max there has been a physically identical stimulus impinging on Yon Max, and for each internal physiological event in Earth Max there has been a counterpart in Yon Max. Their societies, too, are quite similar, though as will emerge shortly, the societies differ in one crucial respect. Both Maxes have long suffered from painful inflammations of the joints, and both have sought treatment for their affliction. In each case the physicians have told them, 'You have arthritis', and their prescriptions were for the same drugs. Earth Max, while hardly an expert on matters medical, does have a modest fund of common lay beliefs about arthritis. He believes that arthritis is not fatal, that stiffening joints is a symptom of arthritis, that aspirin often relieves the pain of arthritis, etc. Yon Max, being a *Doppelgänger*, readily assents to all and only the content sentences which Earth Max would endorse. Now on a certain day Earth Max wakes to find that in addition to his usual painful joints, he also had a quite similar pain in his thigh. He comes to believe that his arthritis has now spread to his thigh, and reports this belief to his doctor. The doctor, however, informs him that this cannot be so, since arthritis is specifically an inflammation of the joints. This comes as news to Max (though he might have learned it from any decent dictionary), but he readily relinquishes his belief and goes on to wonder what might be wrong with his thigh. Yon Max, too, wakes up on the appointed morning with a suspicious pain in his thigh, and he too says to his doctor, 'I think my arthritis has spread to my thigh'. But at this point the lives of Earth Max and Yon Max diverge. For on Yon 'physicians, lexicographers, and informed laymen apply "arthritis" not only to arthritis but to various other rheumatoid ailments.' (Burge, ibid., p. 78.) Thus Yon Max may be right when he says, 'My arthritis has spread to my thigh', since in so saying he is using 'arthritis' with the meaning and extension it has in *his* linguistic community. Now it is Burge's contention that Yon Max lacks 'some—probably all' of the beliefs about arthritis that Earth Max holds. Yon Max lacks the beliefs that he has arthritis, that stiffening joints are symptoms of arthritis, that aspirin often provides relief from the pain of arthritis, etc. Yon Max would, of course, assent to each of these content sentences. But in so doing, Burge maintains, he would be expressing a belief whose content is different from the belief we (and Earth Max) would express using the same sentence.

I think that Burge is clearly right that in the case imagined, and in a number of others that he discusses, there is a strong intuitive pull toward counting the beliefs in question as *different* in content, and thus to deny that Yon Max believes he has arthritis in his thigh. However, I think Burge is wrong in maintaining that the sort of difference in social setting and linguistic practice which he considers is generally *sufficient* to engender the intuition that the beliefs in question are different in content.[18] Rather, it would seem that similarity of social and linguistic setting is one among several factors to be assessed in determining content similarity and in ascribing content. Moreover, the relative importance of social and linguistic similarity in determining our intuitions about content is very much a function of the pragmatic circumstances surrounding the judgement. In the cases Burge elaborates, the difference between the two societies is emphasized, since that difference alone leads to a parting of previously parallel life histories. But it is easy enough to tell similar stories that do not evoke the intuitions Burge would urge. Suppose, for example, that we alter our tale in a single respect. On the fateful morning both Maxes wake up with a new pain not in their thighs but in their elbows, a region hitherto unafflicted. Each acquires a belief which he expresses with the sentence, 'My arthritis is now in my elbow'. And in each case the belief is confirmed by the physicians. Here, I think, the intuition that Yon Max lacks all of Earth Max's beliefs about arthritis is much weaker and more difficult to sustain. This is just what we would expect on the theory of content ascription I have been defending.

My theory also leads us to expect that, keeping pragmatic context constant, greater differences in social and linguistic setting will have greater impact on intuitions about content. And here too the theory seems to fit with the facts. To see this, consider a pair of variations in our story about Yon. On the first variation, the linguistic practice of Yon is rather closer to that on Earth. Physicians and experts on arthritis have adopted a usage for 'arthritis' quite parallel to the one prevailing on Earth. In professional journals and medical 'shop talk' rheumatoid ailments like those afflicting Yon Max would not be labelled as 'arthritis'. But neither the lay dictionaries nor the usage of laymen has caught up with this medical usage. Lay persons generally use 'arthritis' in referring to an ailment like Max's, and doctors adopt this usage also in talking with their patients. In this case, would it be intuitively plausible to say Yon Max believes he has arthritis in his thigh? The answer, I think, is that it would be considerably *more* intuitively plausible than it was in the original version of our tale. Finally, consider a second variation on which Yon physicians, lexicographers and informed lay persons use 'arthritis' in a way which is very different from its use on Earth. Suppose, for example, that Yon physicians, etc. use 'arthritis' to refer to what we would call 'arthritis' when it occurs in the joints of the arms or hands, but not to what we would call 'arthritis' when it occurs in other joints. What is more, informed Yonderfolk use 'arthritis' also to refer to various tumours on the arms and to corns and carbuncles. (Perhaps they have a theory to the effect that all these conditions have a single etiology.) Yon Max, however, is something of a bumpkin and has no clear idea what afflictions apart from his own (and those similar to it) would be called 'arthritis'. How plausible is it, in this case, to say that Max believes he has arthritis in his thigh? It is, I think, considerably *less* plausible than it was in the original version of our story. Indeed, as the divergences between Yonder usage and ours grow greater, we hardly know what to say about the content of Max's beliefs short of describing how his compatriots use their words, and then using *their* language in the content sentence.

VII. Conclusion

My project in this paper was to do a bit of domestic anthropology—to explicate the folk concepts and theories that underlie the practice of ascribing content to the beliefs of our fellows. Plainly much remains to be said on the topic, though I am inclined to think most of it is best said in the language of the cognitive simulator. It remains to ask why the project has been worth pursuing at all. Part of the answer, of course, is that the folk theories and concepts invoked when we ascribe contentful beliefs to people are a basic part of our common-sense world view. So in seeking to understand these theories and concepts we are fulfilling the Socratic edict to know thyself. However, in my own case the principal motive is less venerable. About two decades ago cognitive psychology began an extended flirtation with the concepts of folk psychology. Terms like 'belief', 'memory', 'plan' and others began to find their way into serious psychological theorizing. Nor were these mere homonyms. Though some effort was made to render the terms a bit more 'precise' than they were in ordinary usage, the bulk of our folk concepts and theories were simply adopted wholesale by cognitive theorists. This flirtation has blossomed into a fecund marriage. We now have theories of reasoning, problem solving, inference, perception and memory all cast in the intentional idiom of contentful states. These theories couch their laws and generalizations in terms of the *contents* of the various states they study. Thus, for example, a theory of inference will attempt to specify which beliefs are more or less likely to be inferred from an initial set. The regularities will be specified in terms of the contents of the beliefs (cf., for example, Wason and Johnson-Laird 1972). Many observers have noted that cognitive scientists have never paused to explain what it means to ascribe a content to a belief or a memory, and thus there is a serious lacuna in their theories. But some of us harbour darker suspicions. We fear that the notion of a contentful belief or memory, borrowed from folk psychology, may be singularly unsuited to the purposes of scientific psychology. And we suspect that an analysis of what we mean when we ascribe content to a belief may reveal many of the offspring of the marriage between folk psychology and cognitive science to be lame or stillborn.[19] Nothing I have said in my account of content ascription constitutes an argument for these sombre suspicions. Though if I am right about content ascription, the arguments will not be hard to construct. But that is a project for another paper.[20]

Notes to chapter 4

[1] For an elaboration of the view that analytic philosophy at its best is continuous with cognitive simulation, see Todd (1977); Sloman (1978), ch. 4. For an example of the sort of cognitive simulation I have in mind, see Schank and Abelson (1977).

[2] Much of the story I have to tell about states is modelled on Kim's account of events (cf. Kim 1969, 1976).

[3] The standard source for the opposition on this point is Melden (1961). Dennett (1973) also appears to oppose the assumption. The case for the assumption has been made by many writers including Brandt and Kim (1963), Davidson (1963a), Alston (1966) and Goldman (1970).

[4] For some additional detail, see Lycan forthcoming; Harman (1970); Goldman (1970).

[5] Yet another distinction embedded in our common-sense folk psychology divides beliefs from other contentful, belief-like states which often give rise to beliefs (Stich 1978a).

[6] For more on the distinction between what we believe and what we could quickly

(and unconsciously) infer, see Dennett (1975) and Field (1978). For some cleverly designed experiments designed to determine what is recovered from memory and what is inferred, see Reder (1976).

[7] Cf., for example, Harman (1970); Fodor (1975); Field (1978); Lycan (forthcoming). For a serious attempt to construct a theory along these lines, see Anderson and Bower (1973); Anderson (1976).

[8] This is as good a place as any to confess that I am being systematically cavalier about the niceties of use and mention. Those who are irked by the practice should have no difficulty in paraphrasing my remarks so as to keep careful track of the distinction.

[9] I should note that, while I am borrowing Davidson's idea, I do not share his motivation. Davidson's analysis of *oratio obliqua* is offered with an eye toward providing Tarski-style semantic theories for natural-language sentences. But this is a goal which I take to be quixotic (cf. Stich 1976; Martin 1978).

[10] Davidson himself long ago proposed in conversation that his account of indirect discourse could be extended to belief-sentences. He takes up the point briefly in Davidson (1975), pp. 18 ff.

[11] Quine *aficionados* may be forgiven if they are beset by a sense of *déjà vu*. Consider the following passage:

'In indirect quotation we project ourselves into what, from his remarks and other indications, we imagine the speaker's state of mind to have been, and then we say what, in our language, is natural and relevant for us in the state thus feigned. An indirect quotation we can usually expect to rate only as better or worse, more or less faithful, and we cannot even hope for a strict standard of more and less; what is involved is evaluation, relative to special purposes, of an essentially dramatic act. Correspondingly for the other propositional attitudes, for all of them can be thought of as involving something like quotation of one's own imagined verbal response to an imagined situation.

Casting our real selves thus in unreal roles, we do not generally know how much reality to hold constant. Quandaries arise. But despite them we find ourselves attributing beliefs, wishes, and strivings even to creatures lacking the power of speech, such is our dramatic virtuosity. We project ourselves even into what from his behaviour we imagine a mouse's state of mind to have been, and dramatize it as a belief, wish, or striving, verbalized as seems relevant and natural to us in the state thus feigned.' (Quine 1960, p. 219.)

Much of what I say in this paper can be viewed as elaboration and commentary on this passage.

[12] This view emerges clearly in Davidson (1975), pp. 15–16; it can also be seen lurking in Davidson (1974a), p. 8, and perhaps in Armstrong (1973), pp. 25–6.

[13] I see a clear anticipation of all this in the passage from Quine (1960) quoted in note 11 above.

[14] Haugeland (1979) can be read as arguing that a large number of other dimensions can sometimes influence our content judgements. And while I would quibble over details, I am inclined to think that he is basically correct.

[15] Actually, it would be better to treat 't' as a free variable as well, thus allowing an object to satisfy a functional theory at one time and not at another. But I will ignore this complication.

[16] I failed to note this point in Stich (1979), to the detriment of the argument I was developing.

[17] Lycan, for one, insists: 'Every belief I know of is a belief *that* (a belief that such and such is the case). Were we to report such a belief, therefore, we would do so by tokening

a compound sentence containing a sentential complement; the sentential complement would of course express the content of the belief. Thus, to each belief there corresponds an actual or potential sentential structure. To put essentially the same point in a different way: No doubt there are many sorts of mental processing that trade in structures that are nonsentential. But why would we have any inclination to call any of the relevant processing states *beliefs* in particular?' (Lycan forthcoming, section IV).

[18] Actually, it is less than clear that Burge is interested in explaining our intuitions about content and difference in content. The early pages of his essay contain some suggestive remarks about the role of 'contextually relevant purposes' (1979a, p. 74) in determining what we might count as the same thought or the same belief. But he dismisses these factors as irrelevant to a suitably 'theoretical' notion of content.

[19] Among those who seem to share my suspicions are Churchland (1979), Dennett (1978a) and Haugeland (1978).

[20] The bulk of this paper was written while I was a participant in the Workshop on Philosophy of Psychology held at Bristol University in 1978–9. During my stay in Bristol I was supported by the U.S.–U.K. Educational Commission (Fulbright–Hays) and the American Council of Learned Societies whose assistance is gratefully acknowledged. The ideas in this paper were discussed frequently in the workshop, and most of the members of the workshop helped me clarify my exposition in one way or another. Two workshop colleagues need special mention. Daniel Dennett generated ideas, insights and criticisms with a boundless enthusiasm. And Andrew Woodfield read and reread countless drafts of the essay, providing me with invaluable suggestions each time around. But for his help the paper would be much drearier to read, and would have been much less fun to write. Woodfield was also the principal organizer of the workshop; its success was largely due to his tireless administrative activities. To him, with thanks, the paper is dedicated.

5

The Structure of Content

Colin McGinn

When we report what an agent believes we specify the *content* of his belief (and so for the other propositional attitudes). When we say what a sentence means we similarly specify the content of the sentence. The linguistic form of such content specifications consists in the occurrence of a 'that'-clause within the scope of 'believes' or 'means', the whole being attached as complex predicate to a singular term designating the subject of the belief or the sentence whose meaning is thus specified. The 'that'-clause embeds a (declarative) sentence capable of standing alone, the meaning of which contributes in some systematic way to the meaning of the whole content ascription. A natural view of such constructions is that 'believes' and 'means' serve to express two-place relations—between an agent or sentence and a 'content', whatever a content may turn out to be. These two types of content specification are clearly going to be intimately related, and so it is reasonable to hope that reflection upon each in the light of the other will generate reciprocal illumination. In what follows my project is to investigate what both types of content involve; I want to know, that is, what makes such content ascriptions true. This project will take in a large number of issues, not all (or perhaps any) of which can be fully treated here; my aim is to sketch the general outlines of a position and allude to some consequences for certain doctrines and problems. I shall start by considering some suggestions of Fodor about propositional attitudes; then I move on to meaning.

II

According to Fodor, beliefs (and I shall take these as exemplary) involve relations to internal representations: to believe that p is to be in a certain relation to some internal state s which represents

the objects and properties in the world that the belief is about
(see Fodor 1975, 1978, 1980; Field 1978). Fodor thinks that the
system of such representations is actually a language, so that
believing is structurally like (indirect) saying: the agent stands
in a relation, accepting or uttering, to a token sentence, inner or
outer, which gets interpreted by being assigned certain semantic
properties and relations. I will not myself assume this linguistic
view of the medium of mental representation, but I want to
agree with Fodor that beliefs involve (must involve) relations to
internal representations of *some* sort. Now, granting that
conception of belief, the following question arises: in virtue of
what do beliefs play a role in the agent's psychology? And it
seems, Fodor contends, that there can be only one answer to this
question: beliefs play a role in the agent's psychology just in
virtue of intrinsic properties of the implicated internal representations—the semantic *relations* between representations and
things in the world must be irrelevant to the psychological role
of beliefs. More precisely, the causal role of a belief must depend
upon, and only upon, those properties of representations that can
be characterized without adverting to matters lying outside the
agent's head. Since, as Fodor claims, cognitive psychology is
concerned exactly with the causal–functional role of representations, it takes mental processes and procedures to operate
exclusively upon the intrinsic aspects of belief, and so must ignore
the referential properties of representations. In other words,
cognitive psychology is committed to what Putnam (1975a)
called methodological solipsism. This methodological constraint
thus implies that the theoretical taxonomy of mental states
demanded by cognitive psychology will be determined solely by
such solipsistic features of representations; mental states will be
the same or different according as their causal role is the same or
different.

Now typically semantic distinctions between mental representations are mirrored in intrinsic differences between them: what
the corresponding beliefs are about is normally, as we might say,
encoded within the internal representations—or equivalently, the
solipsistic properties of representations *determine* their semantic
characterizations. When there is such encoding the taxonomy of

states required by a psychology of the causal role of propositional attitudes matches the taxonomy we get by allowing what the attitude is about to contribute toward its individuation. But as Fodor observes, drawing upon recent discussions of reference and belief, such matching is not invariably the case. He mentions two sorts of example. First there are cases in which beliefs are formed about different things in 'qualitatively indistinguishable' circumstances—what have come to be called Twin-Earth cases.[1] Here it has been argued that the beliefs cannot be the same, if beliefs are individuated by the meaning of the sentence that specifies their content, since the content of a belief is (in part) fixed by what the embedded content sentence says the belief is about—and in Twin-Earth cases the beliefs are about different things and so may differ even in truth-value. So we seem to have cases in which the states of the head that function as internal representations do not suffice to fix the full content of the belief, because in these cases the semantic relations are *not* intrinsically encoded. We need to appeal to causality or context or some such to determine what the belief is semantically about, and hence what its truth-conditions are. The second kind of case is that of indexicals (cf. Kaplan forthcoming; Perry 1977, 1979; Lewis 1979). It is a commonplace (though an important one) that indexical beliefs may be ascribed by content sentences with the same linguistic meaning (character, in Kaplan's terminology) but differing in referential truth-conditions. For example, two people may self-ascribe some property using the word 'I' and the ascriptions differ in truth-value, even though the internal states of the people are themselves indistinguishable; so plainly such internal states do not on their own determine the truth-conditions of the expressed beliefs. (Here we can say that reference is determined by the occurrence of a representation *in* a context, not by way of a representation *of* the context. On the relation between concept, context and referent, see also Burge 1977, 1979b). In both kinds of case it appears that the intrinsic properties of the representations fail to determine the semantic content of the beliefs in which the representations figure. It follows that the causal role taxonomy delivered by methodological solipsism does not match the taxonomy suggested by ordinary content ascriptions, for which the identity of a belief is (partly) a matter of its truth-conditions. In short, the truth-conditions of a belief are not always or necessarily encoded in its causal role. But this should come as no great surprise, in view of the structure of belief, since it was already clear that it is the representations *themselves* and not their referential properties that play a causal role in the agent's psychology: even where reference *is* intrinsically encoded it is not the *relation* to the referent that is causally relevant but the way in which the referent is represented internally. It is not *what* is encoded that matters to causal role, but what it is coded *into*. So even if there were a one–one mapping between representations and referents the basic point would remain: semantic properties are irrelevant to a psychology of causal role; at best they figure by proxy. On the other hand, it seems undeniable that beliefs do have (something like) semantic properties—for they have referential truth-conditions—and these properties do exert a pull in the individuation of beliefs by content. It thus appears to follow that ordinary content ascriptions, made as part of commonsense psychology, are not methodologically solipsist. What Fodor suggests is that we need *two* kinds of psychology to deal with propositional attitudes: a psychology of internal representations, which is methodologically solipsist; and a psychology of the representation relation, studying the extrinsic relations between representations and the environment.[2] Ordinary content ascriptions conflate these two aspects of belief.

I think we should conclude from these observations that our intuitive conception of belief-content combines two separable components, answering to two distinct interests we have in ascriptions of belief. One component consists in a mode of representation of things in the world; the other concerns itself with properly semantic relations between such representations and the things represented. I want to suggest that the former component is constitutive of the causal–explanatory role of belief, while the latter is bound up in our taking beliefs as bearers of truth. We view beliefs *both* as states of the head explanatory of behaviour, and as items possessed of referential truth-conditions (cf. Perry 1979). (My talk of 'referential' here is loose. I qualify it later.) These components and the concerns they reflect are distinct and independent—total content supervenes on both taken together. We get different and potentially conflicting standards of individuation—and hence different conceptions of what a belief essentially is—according as we concentrate on one or other component of content. The tendency of discussions of belief is, I think, to allow one component to eclipse the other, thus producing needless conundrums and a distorted picture of the nature of belief. We can best see our way through problems about belief if we recognize that belief-content is inherently a hybrid of conceptually disparate elements, *both* of which inform our conception of belief and its individuation. On this dual component view we will be prepared to expect—precisely what we find—that beliefs may have the same truth-conditions and different explanatory role, and the same explanatory role accompanied by different truth-conditions. Words in content clauses—singular terms or predicates—may thus be said to make a dual contribution (Loar 1972; Schiffer 1977) to the truth-conditions of the whole content ascription: they indicate the character of the representations the agent employs in his thought about the world, *and* they specify which objects and properties are relevant to the truth-conditions of the belief itself. Neither contribution suffices to determine the other, and both are needed to fix total content.

Starting from Fodor's idea that belief involves internal representations, I was led to advocate a dual component conception of the structure of belief-content. Now I want to try to reach the same conclusion on the basis of the very nature of representation: representations, I want to argue, necessarily have two aspects and these two aspects must be mutually independent. Let us first ask what is the point or function of representations. The answer, I take it, is that representations are made to mediate between the agent and his environment in such a way as to permit or produce appropriate action guided by information about the world. That is, a mental representation is a state whose function and *raison d'être* is to control behaviour in the light of evidence; a representation, we can say, is precisely that which discharges this role. Thus a perceptual experience, for example, can be said to represent the environment, and the point of its so doing is to enable the perceiver to act appropriately in respect of the perceptually represented environment. So any *theory* of mental representation should, I think, address itself to the role of representations in the organization of behaviour. On the other hand, representations must also have properties *beyond* those constitutive of their intra-individual causal role, because they at least purport to relate to extra-individual entities and states of affairs—they have a 'referential' aspect. To claim that semantic properties are invariably encoded in intrinsic properties is then to hold that the intra-individual role of a representation determines its extrinsic relations; or in other words, that reference is somehow supervenient upon dispositions to behaviour. However, I think this supervenience claim conflicts with two important, perhaps definitive, properties of representations. One is that representations can perform their function even when they *incompletely* or *improperly* represent the object or state of affairs in question. This is because a creature can, so to speak, rely on its spatio-temporal context to resolve ambiguities of reference, and thus ensure that the behaviour caused by the representation is appropriate to the object represented. Thus a perceptual experience can direct a creature's action to the right object in its environment despite the fact that there is some other object in the world which fits the experience equally well— Twin-Earth cases well illustrate this point. But second, and more important, it seems to me to be part of the concept of representation that if r represents x as F it is (epistemically) possible that x not be F—i.e. representations are necessarily *fallible*. How an object is represented should therefore be seen as inherently vulnerable to failures of verisimilitude; it cannot be that some state of a creature should qualify as a representation and yet be logically guaranteed to represent reality correctly (unless, of course, the reality represented is of a special sort). It is therefore a condition of adequacy upon a theory of mental representation that it render representations fallible. But now it seems that you can fallibly represent an object as F only if *what* is represented is fixed independently of its being F. For suppose the fixing were not thus independent. Then error in the representation would entail that the representation was not really *of* the object in question—it would be a representation of some other object that did fit it or of no determinate object at all if the representation were improper. In order for us intelligibly to judge that *r* *mis*represents *x* as *F* we have to assume that *r*'s being a representation of *x* is independent of its characterization of *x* as *F*. This rather abstract point can be appreciated from the nature of perceiving and naming. A visual experience will represent the object of perception in a certain way—the object will be seen *as* such and such. To regard the experience as genuinely a representation one has to admit the possibility that the object is not in fact as it is seen to be—the experience fallibly represents the object. Such fallibility presupposes that the erroneous experience is still *of* that which it erroneously represents. And of course we do suppose that an experience can be a perception of an object which it in fact misrepresents. This is possible because the perception relation is fixed independently of the mode of perceptual representation; it is fixed by a certain kind of causal relation. In other words, the intra-individual role of the experience does not determine its extrinsic relation to the perceived object, precisely because of its representational fallibility. With respect to proper names we can also distinguish two aspects: there is the role of the name in thought and action, and there is the relation of reference between the name and its bearer. Does the former determine the latter? Inasmuch as a name is (or is associated with) a mental representation I do not think it could; for we want to make sense of the idea that the user of the name might misrepresent its bearer to himself. I think we can view Kripke's case against description theories of naming in this way (though he himself does not present his counterexamples in these terms in Kripke 1972). The descriptions a speaker associates with a name constitute his representation of its bearer; but these, Kripke argues, may fail to fit the bearer; so the naming relation cannot be *set up* by such representing descriptions. There could only be such cases of erroneous descriptive beliefs if the naming relation were fixed independently of the representing descriptions. If I am right that representations are essentially fallible, the existence of such cases can be predicted just from the idea that names involve representations: for we may misrepresent that of which we speak. The fundamental mistake of description theories is, then, that they make linguistic representation by names necessarily infallible; for a name will simply not, on that theory, have an object as bearer if the associated descriptive beliefs are not true of it, being perhaps true of some other object or of no unique object. So names must have a component of their content which attaches to them independently of *how* the name represents the denoted object; the two aspects of the name are thus independent. Looked at in this way there was nothing adventitious about the availability of the examples exploited by Kripke against the description theorist. I therefore think that the incompleteness and fallibility characteristic of representations give the lie to the thesis that the action-guiding intra-individual role of a representation can determine its referential aspect. But then if representations do of necessity have these two aspects independently, their content will *eo ipso* comprise two separable components—*what* is represented and *how* it is represented. These, of course, are just the components we earlier isolated upon consideration of Fodor's conception of belief, which also seemed forced upon us by Twin-Earth examples and indexicals, and which are specified by way of the special dual contribution made by words in belief-contexts. Representations being what they are, content has the structure it has.

Fodor suggested that the two components of belief be the objects of two different kinds of psychology; both together would tell us all there is to know about belief. But commonsense psychology—in particular, the rationalization of action by citing desires and beliefs—does not itself break belief-content down into the two components; its explanatory predicates appear to invoke properties that straddle Fodor's two kinds of psychology. It can easily seem that there is a problem about this: it can seem that either it is, after all, wrong to restrict the explanatory role of beliefs to their internal component; or that ordinary belief–desire psychology cannot be a causally explanatory theory, since it violates methodological solipsism in respect of the properties it employs in rationalizations. That is, there seems to be a tension between the causal–explanatory pretensions of commonsense rationalization and the conditions on what sort of psychological property *can* be causally explanatory (cf. Stich 1978b). However, I think there is no real inconsistency here. To see this, consider factive propositional attitudes, e.g. knowing, remembering, perceiving. We do commonly employ these in explanatory contexts, yet it would be agreed that they are hybrid states requiring the world as well as the agent's head to be a certain way. What we should say of this is clear: only the internal component of the condition reported is doing explanatory work—the rest is, from an explanatory point of view, idle. Since I similarly hold that beliefs have hybrid content I can take a similar line: the explanatory force of the content ascription attaches only to the contribution the words in the content clause make in their capacity as specifiers of internal representations; their referential properties play no explanatory role. Indeed, one might see factives and propositional attitudes in general as instantiating a common feature: both require the embedded sentence to have certain *extensional* properties—that the sentence be true or that the terms and predicates occurring therein should succeed in referring. So we can say that the truth-conditions of content ascriptions are indeed not solipsistically specifiable, but only *part* of the truth-conditions is relevant to the explanatory role of the belief. This does, of course, have the consequence that not every aspect of rationalization can be assimilated to causal explanation, since a component of content is irrelevant to causal role.[3] Commonsense psychology is not methodologically solipsist in the properties it invokes, but we need not conclude that it violates methodological solipsism with respect to its explanatory dimension. There is however the question, once this has been granted, as to whether the genuinely explanatory states are themselves beliefs of some sort, i.e. whether we can ascribe those purely explanatory states by way of a content specification. To claim that the explanatory states are beliefs would be to claim that underlying any nonsolipsistic ascription of belief there is a purely solipsistic belief. I see no good reason to accept this, either with respect to singular terms or predicates in content clauses. It amounts to the idea that there can be purely 'qualitative' belief-contents; I doubt that we have any such beliefs and I do not know what their ascription would look like (cf. Kripke 1979, p. 262). If this is right, it is as misleading to say that beliefs and desires causally explain action as to say that factive attitudes do. It is just that we ascribe such nonpropositional causally explanatory states *by* ascribing genuine propositional attitudes.

The thesis that content has both explanatory and truth-related components throws some light on the difference between transparent and opaque ascriptions of belief. When a term occurs transparently outside the belief-context it specifies only the truth-related aspect of content: this is shown in the fact we are thereby enabled to evaluate the belief for truth, but we cannot use the transparent ascription to explain the agent's actions, for this requires us to know how the agent represents the objects of his belief. When a term occurs opaquely within the belief-context it gives us two sorts of information: the truth-conditons of the belief, and the representations that causally explain the agent's actions. If it is asked why we do not have belief-sentences which ascribe *only* representations, I think the answer is that beliefs are essentially bearers of truth-conditions; so we need some way of getting from the truth of a belief ascription to the truth-conditions of what is believed—we need a route to reference— and representations on their own do not supply such a route.

My thesis, to summarize, is that our concept of belief combines two separate elements, serving separate concerns: we view beliefs as causally explanatory states of the head whose semantic properties are, from that point of view, as may be; and we view beliefs as relations to propositions that can be assigned referential truth-conditions, and so point outward to the world. This bifurcation of content can be seen as stemming from the point that beliefs involve internal representations, and these inherently present a dual aspect. This thesis will be confirmed and clarified by an examination of the structure of *sentence* content, to which I now turn.

III

As I remarked at the outset, the logical form of ascriptions of meaning to sentences seems very like that of ascriptions of content to beliefs: but is there any closer tie than that? If we can show the two types of content to be linked in some way, then considerations about meaning will tell us something about belief-content and vice versa. I will mention three sorts of link that might be suggested. First, we have those theories of belief-sentences which make belief a relation to the sentence embedded in the content clause—as with quotational and paratactic theories (Davidson 1968, McDowell 1980). What a person believes, on such theories, is just the content-specifying sentence (token or type); so belief-content turns out to be the same as sentence content. Second, there is the language of thought hypothesis: according to it, belief is a relation to an *internal* sentence, which need not be a sentence either of the ascriber's (public) language or that of the believer. To give a semantic theory of the content of the internal language would therefore be to give a theory of belief-content (Field 1978). But third there is a less theoretically committed way of forging a connection: since the meaning of the embedded sentence serves precisely to specify the content of the ascribed belief, it seems right to identify contents with meanings. That is, belief is a relation to a sentence meaning, though not necessarily a sentence the believer understands. So when a person expresses a belief by uttering a sentence, we can say that he believes precisely what his utterance means. This identification is, of course, close to Frege's view of oblique contexts: what a person believes is the thought (sense) expressed by the obliquely occurring sentence (Frege 1892). Let us put this by saying that that which is grasped by someone who knows the meaning of a sentence is the same as that which is believed by someone whose belief is specified by using that sentence. Given this identification, conclusions reached in one area will be directly transferable to the other; and I think we shall see that the issues that arise in respect of meaning confirm this identification.

I said that we treat belief-content in two ways: as explanatory and as truth-evaluable. Let us analogously ask what the notion of sentence meaning is designed to do for us—why do we have the notion? When we know to what purposes we put the notion of meaning we shall be better able to see how meaning should be theorized. In other words, we are to think of meaning as a theoretically introduced concept and then ask what is the nature of the theory in which it figures. One very influential and appealing answer is that the concept of meaning is tied to the explanation of the *use* of language—what we *do* with sentences. A closely related answer is that the notion of meaning has its theoretical role in a characterization of the cognitive states which constitute the *apprehension* of meaning. Consequently any adequate account of use or understanding would *be* a complete theory of meaning; put differently, any set of concepts introduced to explicate the nature of meaning must be adequate to explain use. The positive aspect of the use conception of meaning is that a theory of meaning must *at least* explain use; the negative aspect is that the theory must *at most* explain use. So the question becomes: what feature of sentences determines their use?— meaning will then consist in that use-determining feature. Three writers in whom this point of view is more or less explicit are Dummett (1976, 1978), Putnam (1978, notably in 'Reference and Understanding') and Harman (1973, 1974). Each of these writers, in their different ways, takes the use conception to rule out the introduction of certain concepts in a theory of meaning and to invite the employment of others. Thus Dummett disputes the employment of the (classical) notion of truth in the characterization of sentence content and favours notions of verification and falsification; Putnam extrudes reference and truth from the theory of understanding and use and invokes notions of acceptance and subjective probability in their stead; and Harman rejects a truth-conditions theory of meaning, proposing a functionalist theory. Each of these proposals carries a certain conception of what the state of semantic understanding consists in: that state will be defined by the dispositions to verbal (and other) behaviour it induces. What determines use is a state of the head—'knowing meaning'—and this state gets character- ized in terms of the behavioural manifestations in which it is displayed. I hope it is not too tendentious to say that, according to the use conception of meaning, as exemplified in these writers, meaning is a matter of the causal role of the state of semantic understanding.

It should be clear enough that such a conception implies or

assumes methodological solipsism in the theory of meaning, since what explains use is precisely a state of the head. That is why *reference* seems to drop out of theories of meaning shaped by the use conception.[4] We should be reminded here of an analogous result concerning belief-content: when beliefs are viewed exclusively in their capacity as explanatory states the truth-conditions component of content has no place. Think of it this way: understanding is a relation to a content, as is belief; but if we insist on viewing these attitudes as exhausted by their explanatory role it will seem that the referential component of content is idle—there will be theoretical room only for the internal representations. Meaning, on the use conception, comes to be a matter of what I shall call *cognitive role*—and this is an entirely intra-individual property. Again, the in principle irrelevance of semantic relations to explanatory role is obscured by the fact that, very often, cognitive role determines reference. But, as we saw with belief, what is in the head does not always or necessarily determine reference;[5] and even when it does it is the cognitive representation itself and not what is coded into it that is strictly constitutive of explanatory role. Once we give up the idea that reference is determined by what is in the head—by cognitive role—it becomes clear that a theory of meaning conforming to the use conception will find no legitimate place for referential truth-conditions.

But I do not think we can rest content with a pure cognitive role theory, discarding reference and truth entirely. For any adequate theory of language should address itself to relations between words and the world, since such relations are clearly (to say the least) a very important feature of language: we cannot just choose to ignore semantic relations in our theory of the workings of language. Furthermore, there seems good reason to suppose that reference and meaning are not independent properties of expressions: the meaning of a sentence surely determines its truth-conditions, and one who did not know the truth-conditions of a sentence would not fully grasp its meaning. The difficulty is that it does not seem possible to motivate reference on the use conception. So how could we warrant its introduction into a theory of meaning? In my discussion of belief

I said that reference comes in by way of truth; and this does motivate it in a conditional way, since truth requires reference. But this does not answer a more radical question, pressed by Hartry Field (1972, section V, 1978, section V): what is the basis, if any, of our customary ascription of *truth*-conditions to beliefs and sentences? The question is pressing because the most obvious place to seek the motivation for speaking of semantic correspondence is in the explanatory utility of belief and meaning—but closer scrutiny has shown that place to be already adequately filled by representations or cognitive role. I shall now critically review some suggested answers to this question; to answer it would be to disclose the point of the nonexplanatory component of content.

(i)   Let us first put aside a misguided answer to our question. Frege, having argued that words have sense (cognitive value) as well as reference, asks why it is that we do not confine ourselves to the sense and let the reference go (Frege 1892, p. 63). His reply, in effect, is that we need reference because we are interested in how the world is—we seek knowledge of the world. It is important to see that this cannot be taken as an answer to *our* question. It can seem to be because of an ambiguity in the question 'why should we be interested in reference?' We must distinguish between: why we should be concerned with the *referents* of our words (or better: with respect to our words, why we should be concerned with their referents), and why we should be interested in the reference *relation* between words and things. Frege is answering the first question, but this does not suffice to answer the second. To suppose it did answer our question would be as wrong as trying to refute a redundancy theory of truth by insisting that the concept of truth is indispensable because we want to know the *truth* about the world. To motivate a concern with the world is not to motivate a concern with the *relation* between the world and language. (This is not to deny that an interest in the world might somehow underlie our ascriptions of reference—only that the present suggestion does not on its own begin to answer our question about semantic correspondence.)

(ii)   We can also deal briefly with what Dummett sometimes

suggests is the role of reference in an account of meaning (Dummett 1973, especially ch. 5). Dummett takes the mode of *presentation* (cognitive value) associated with a term to coincide with its mode of *designation* (mechanism of reference): i.e. he holds that the manner of association between word and object is fully mentally represented. He can then claim that the reference relation enters the characterization of meaning as use because, to put it my way, that relation is encoded as an intrinsic property of the speaker which serves to determine his linguistic and other behaviour. From this perspective Dummett can insist that a causal theory of reference, for which the reference relation is not mentally represented, has the undesirable consequence that reference becomes 'idle in the theory of meaning'.[6] In other words, only something like a description theory of reference—which locates the mechanism of reference within the head—can justify the semantic relevance of reference; but granted that theory, reference *does* play a role in a use-centred theory of meaning, it being precisely that which sense (cognitive value) determines. This way of introducing semantic correspondence cannot satisfy us for two reasons, both of which have come up before. First, it is not true that what is in the head in the way of mental representations suffices to determine reference—we need to appeal to causal or contextual factors. Second, Dummett's suggestion motivates reference only derivatively or by proxy: it says that the two-place nonsolipsistic reference relation enters into the theory of meaning by being coded into a one-place solipsistic condition of the speaker's head. But, by that very fact, the reference relation proper falls away as redundant; it does nothing not already done by what it is coded into (empty terms bear out this point). In this respect, I think Putnam (op. cit. in 1978) has a clearer appreciation than Dummett of the consequences for reference of the use conception: its commitment to methodological solipsism precludes recognition of semantic correspondence.

(iii)   Putnam begins by connecting understanding with use, sees that reference is otiose in a theory of use, and concludes that the learning and mastery of a language—one's knowledge of

meaning—do not involve referential concepts: you can know
the meaning of a sentence without so much as having the concept
of truth. The ability to employ language no more presupposes a
grasp of referential concepts than the ability to turn on lights
presupposes a grasp of the concept of electricity. The concept of
truth, for Putnam, enters at an altogether different level: it
should be construed as a theoretical term in an explanatory
theory of the *success* of linguistic behaviour, as the concept of
electricity is invoked to explain the success of flipping light
switches (1978, p. 99). Elsewhere (1975a) Putnam proposes to
decompose meaning into 'stereotype' plus extension: combining
this with his account of understanding and truth, we could say
that, for Putnam, it is the stereotype that explains use and is that
in which *knowledge* of meaning consists; while the role of the
extensional component of meaning is to account for why the use
behaviour prompted by the stereotype is generally successful.
Putnam's story here is not totally free of obscurity, but it seems
clear enough to invite the following objections. First, it is not
obvious to me that we *require* truth and reference to explain the
fact (if it is one) that behaviour directed by mastery of language
tends to fulfilment of our goals. Instead of accounting for such
success by saying that our utterances and beliefs are reliably *true*
(where the theory of reliability will invoke relations at least
coextensive with referential relations), why not tie success to
evidence? Our actions tend to fulfil our goals because they are
(typically) informed by good evidence, i.e. our utterances tend
to be *assertible*. And if our having reliable evidence can explain
success, it looks as if we have still not found a feature of language
whose explication requires us to go beyond narrow psychology.
Nor is it at all clear that an intuitionist or formalist about
mathematics cannot explain the utility of mathematics. But a
more fundamental and frontal objection is that Putnam's story
unacceptably dissociates one's knowledge of meaning from one's
being able to take assertoric utterances as (primarily) reports on
the condition of the world. That is, an audience interprets an
assertion, not just as the expression of the internal stereotype the
speaker associates with his words, but as standing in a certain sort
of relation to the world; and it seems that you could acquire

knowledge of the world by this means only if you (at some level) took the sentence uttered to have referential truth-conditions, i.e. to be such that the world is a certain way if and only if the sentence is *true*. Surely it is absurd to suppose that someone could achieve full mastery of the practice of speaking a language, using it in communication, and not, in some way, take words as possessed of reference—such a one could not appreciate the point and purport of an assertion. (I come back to reference and communication later.)

(iv)   A quite different suggestion is that we need reference and truth because the only way in which the cognitive role of words and sentences can be *specified* is by citing their reference or truth-conditions—we can *show* the cognitive role of a word only by *saying* what its reference is.[7] We should distinguish this thesis from the claim that we cannot specify *meaning* except by stating reference: that claim seems to me true, simply because reference *is* an ingredient in meaning, but it does not follow that it is the cognitive role of the word that is thereby specified. The confusion here is abetted by uncritical use of the notion of sense, construed as that which constitutes cognitive value *and* that which determines reference, i.e. as roughly equivalent to the intuitive undifferentiated notion of meaning. But we have seen that these two jobs are not and could not be discharged by the same property of an expression. So to say that the specification of meaning requires the use of 'refers' does not yet address the question at issue, since that question just is how the noncognitive component of meaning is to be introduced. Once we are clear that it is the cognitive role that has to be referentially specifiable, I think the suggestion loses whatever plausibility it may have had. Note to begin with that it is of course not *sufficient* for the specification of the cognitive role of an expression that we state its reference; it could not be, since 'refers' produces a transparent context (it is a genuine relation). So we would need some further condition to select such ways of stating reference as would fix cognitive role—i.e. which used expressions have the *same* cognitive role as the mentioned expressions. This clearly calls for some prior criterion of sameness of cognitive role. But still it may be said that this prior criterion cannot be directly exploited to specify cognitive role; it can serve only as a constraint on reference assignments. However, this seems to me very implausible, for mental representations and the things they represent are evidently distinct existences; so how could it be that the former are not identifiable independently of the latter? Consider empty names: it seems to me undeniable that they can have a cognitive role (which is *not* to say that they have what we would ordinarily regard as meaning); but their cognitive role would be ineffable if the present suggestion were right. But we need not rely on these general reasons for supposing that nonreferential specifications of cognitive role must exist, because we do have ways of alluding to cognitive role—whether or not these are ultimately theoretically adequate—which do not employ the concept of reference. Thus the cognitive role of a name might be given by a description—proper or improper, correct or erroneous—either by quoting the description or by availing ourselves of the traditional (though logically rather obscure) device of italicizing the description in order to designate a concept. Or we might, following a suggestion of Dummett's (1973, p. 227), specify cognitive role by saying what *ability* its possession confers on the speaker; or again a functional definition of the dispositions induced by associating a given cognitive role with an expression might be given. It is also conceivable that an adequate account of cognitive role can only be expected from scientific cognitive psychology, in terms perhaps of some kind of subconscious system of representation; so that our usual ways of indicating cognitive role relate to this underlying system in somewhat the way in which our commonsense terms for physical dispositions relate to a properly scientific account of the basis of the disposition. I will return to what a theory of cognitive role might look like later; for now I want to conclude that, since it is not *necessary* (and not sufficient) to use reference to specify cognitive role we are still wanting an answer to our question. It would indeed be extremely puzzling if reference did come in this way, since cognitive role is a matter of use and reference is irrelevant to use.

(v)   Field (1972, 1978) offers an answer to our question which seems to me close to the truth. He wishes to find a serious point in our ascribing semantic properties to sentences and beliefs, given that such properties play no part in a psychology of internal representations. He suggests that we employ semantic concepts because we take people's utterances and beliefs as generally reliable indicators of the world, that this vastly increases our ability to acquire knowledge about the world, and that an articulate theory of how we acquire knowledge in this way will make essential use of the notions of truth and reference. It is important to note that this suggestion locates the point of semantic concepts quite outside of psychological explanation: semantic concepts direct our concern to the world, not to the speaker's behaviour. Note also that the suggestion tacitly connects reference with meaning—with the information conveyed in an assertion—since what state of affairs a sentence represents is plainly part of its meaning: so, unlike Putnam's suggestion, this does make grasp of semantic concepts constitutive of linguistic competence. Thus in Twin-Earth cases, for example, we assign distinct references to expressions because the sentences uttered convey information about different things, and we want to know about those things. (No doubt this practice is founded on a largely implicit theory of learning and reliability; to make that theory explicit would be to say what governs our assignments of reference.) However, as stated, Field's suggestion has some untoward consequences. The suggestion has two parts: (a) we want knowledge of the world (for obvious reasons), and (b) we take it that people's utterances and beliefs are reliable ways of getting such knowledge. But it seems that neither of these considerations could be constitutive of why we assign reference as we do, since they are in a certain sense merely contingent. Suppose that with respect to certain subject matters my knowledge is complete (or I suppose it to be); then I shall not, according to Field's suggestion, have any motive for assigning reference to those of your utterances and beliefs which relate to those subject matters—omniscience has no need of reference. Or suppose that your utterances are *un*reliable (or I take them to be); then again I shall lack any motive for assigning reference to your words. But surely we want to say that a hearer in these relations to a speaker's words *would* still think it proper to assign reference to them; so Field's suggestion seems not to be quite right. The basic trouble is that he makes the *raison d'être* of reference and truth too dependent upon conditions of the *hearer*, instead of locating their point in the characterization of the activity of the *speaker*. We need to associate reference with the very nature of communication.

(vi) Let us then modify Field's suggestion in the way indicated. What seemed right in his suggestion was the idea that reference and truth are needed in an account of how it is that utterances can transmit knowledge about the world. I accordingly propose that we locate reference in the point of communication—in the intention with which assertions (and other kinds of speech act) are made. A hearer understands a speech act as an assertion just if he interprets it as performed with a certain point or intention— viz. to convey information about the world. On this view, omniscience and unreliability will not destroy the motive for assigning reference, since we will still need semantic concepts to explain why it is that the speaker chose *those* words to convey knowledge (or belief) about a certain state of affairs to an audience. Of course the hearer will often acquire knowledge of the world by witnessing assertions, but his doing so is not the root reason for assigning semantic properties to the speaker's words (indeed it is plausible to see our acquiring knowledge in this way as resting upon a prior interpretation of the utterance as performed with an intention whose possession requires a tacit grasp of semantic notions). To fulfil the intention to communicate knowledge (or belief) about the world in language the speaker must exploit signs standing in representational relations to things in the world. In fact, I think that finding a place for reference forces us to take a world-directed view of communication; for if, as some philosophers have held[8] (or at least said), the point of communication is to convey the solipsistic contents of the speaker's mind, then the notion of reference would not be required in characterizing the activity of communicating. But if this *is* the right way to introduce reference, it follows that communication is aimed rather at the world. (It should be noted that my suggestion is not that the reference relation itself explains a speaker's acts of communication; I am saying rather that the *thought* of it is presupposed in reflective mastery of the practice, or in characterizing its function where the practice seems quite unreflective.)

This suggestion may well appear rather obvious once it is formulated, but in fact it creates a prima facie problem. Field's suggested motivation applied equally to utterances and beliefs, indeed fundamentally to beliefs; my amendment is stated in terms of the point of utterances, and so does not seem applicable to beliefs at all, since belief-states are not actions performed with a point. It may then look as if my suggestion has not really uncovered the theoretical root of reference. However, I think I can turn this apparent lacuna to my advantage; for I doubt that beliefs do have genuine referential properties in the manner of sentences—a belief's being *about* something is not the same as a sentence's *referring* to something. (If the aboutness of beliefs were not thus literally referential, then that would cast some doubt on the view that the representations implicated in belief are, properly speaking, sentences of a language, since that view implies a quite literal sense in which beliefs involve reference.) That aboutness is not the same as reference is suggested by the following considerations. First, we are reluctant to apply strict semantic concepts to beliefs in the ordinary course of things; we naturally speak of the aboutness of beliefs as *quasi*-semantical, thus signalling a derivative or analogical usage. Second, part of our reluctance to attribute beliefs to nonlinguistic creatures stems from doubts about assigning a specific and discriminated object to the belief, i.e. about giving it determinate truth-conditions; we tend to let the explanatory aspect of belief dominate in such cases, and feel unsure about judgements of truth-value (cf. Stich 1979). Inasmuch as we wish to attribute beliefs in the case of nonlinguals, their aboutness seems closer to the ofness of perceptual experiences—which no one would directly assimilate to a genuinely semantic relation. It seems to me that the aboutness of beliefs is situated midway between the ofness of perceptual representations and the reference proper of words; so we should not *want* it to be straightforwardly true that beliefs refer. This intermediate status of belief representation is perhaps not surprising in view of the location of beliefs in the causal network of psychological processes: beliefs are (typically) states caused by perceptions and subsequently expressed in speech; so their causal role with respect to input and output seems to reflect (perhaps to underlie) the conceptual status of their representational properties. The aboutness of belief stands between the more primitive (perceptual) and the more sophisticated (linguistic) ways of representing the world, and it reflects a bit of both. Suspended between perception and reference beliefs can be pulled now in one direction, now in the other: when language is absent (so-called) beliefs approximate to perceptual states; when language is present it is apt to dominate and make us conceive of beliefs as essentially linguistic in character. But now we still have the question of why we ascribe even nonreferential aboutness to beliefs. Part of the answer is that we think of beliefs as essentially communicable, so that some of the referential quality of words rubs off on them. In so far as we do not conceive of belief linguistically I think the answer will coincide with the correct answer to this question: why do we say that perceptual experiences are *of* things in the world? I suspect that we do so because of the phenomenon of *learning*: perception is a process whereby a creature acquires information about its environment, and learning in the most basic case consists in the reception (or registration) of sensory stimuli caused by the perceived object. Perception is of objects because it is the means by which creatures learn about things around them; we account for such learning by postulating relations of perception—where the resulting cognitive states guide behaviour in respect of objects in the environment.[9] No doubt this account of the theoretical role of the notion of perception is crude and sketchy, but I think it does something to explain why we say beliefs are about things and what such aboutness consists in. The representational properties of beliefs are a sort of combination of properly semantic reference and perceptual ofness.

I started this section by asking after the theoretical role of meaning. I agreed with those writers who tie meaning to use, and introduced the idea of cognitive role as what determines use. But I did not agree that meaning is exhausted by use, since this leaves out reference. I then motivated reference in a different way, as the means by which language conveys the condition of the world. It emerges, then, that the notion of meaning, like that of belief-content, is structurally duplex: it comprises two distinct components, each component introduced to serve a different purpose and each to be theorized in conceptually different ways. In the next section I shall try to articulate further what a complete theory of meaning, shaped by the preceding reflections, would be like.

IV

In the case of belief we said that the two components of content are specified by way of a dual contribution made by expressions in the belief-context—they attribute a representation and they relate that representation to things in the world. We can, similarly, construe words in the 'means that' context as making a dual contribution of the same sort—they specify the cognitive role of the mentioned expressions and they specify their referential properties. For perspicuity we can separate out the two contributions by taking the meaning ascription as equivalent to a conjunction: for s to mean that p is for s to be true iff q, for some 'q' having the same truth-conditions as 'p', *and* for s to have some cognitive role φ such that 'p' also has cognitive role φ. (Notice that no position in this conjunction is nonextensional; the intensionality of 'means that' results from compounding the two contributions.) Now to have a complete theory of meaning would be to have adequate theories corresponding to each conjunct of this schema; the ordinary style of meaning ascription could then be seen as combining or straddling these two theories (neither of which would directly employ the concept of meaning), in much the way that we saw the ordinary notion of belief-content to straddle the two sorts of psychological theory envisaged by Fodor. But it seems that nothing of critical importance would be lost, and some philosophical clarity gained, if we were to replace, in our theory of meaning, the ordinary undifferentiated notion of content by the separate and distinct components exhibited by the conjunctive paraphrase. An analogy here might be this: in ordinary talk we employ the concept of a reason, but the ascription of a reason really consists of a conjunctive ascription of a belief and a desire in combination. The concepts of belief and desire are quite distinct, and a particular belief–desire pair combining to form a reason can be detached and combined with other desires and beliefs to form different reasons—so clearly there is no mutual determination as between the beliefs and desires that jointly constitute reasons. In roughly this way I think that reference and cognitive role go to make up the intuitive idea of meaning: they too are conceptually disparate and mutually independent, permitting variation of pairings. Let me now mention three general consequences of this picture of meaning, before delineating the precise shape of each sub-theory.

The first is simply that the dual component conception appears to conflict with an assumption (or thesis) of Dummett's about the proper form of a theory of meaning, namely that it will employ a single central concept in terms of which all of the traits of meaning will be explained (see e.g. Dummett 1973, pp. 360f, 1976, pp. 75f). It may seem slightly odd that Dummett should hold this view, since he also insists upon a firm distinction between sense and reference; but there is no inconsistency for Dummett here, because he also holds both that sense determines reference and that sense is to be characterized in terms of referential concepts. We have seen, however, that this view of sense is fundamentally wrong, if sense is understood as cognitive role or mental representation. Once the intra-individual role of expressions and their reference fall thus apart any unity in the notion of meaning dissolves, and we find ourselves accounting for each component by means of distinct sets of theoretical terms (as below). Nor does this duality seem objectionable when the two dimensions of meaning are properly distinguished: not all of meaning is truth-conditions and not all of meaning is use, but what is not either of these is not of the same *sort* as what is. So a verification conditions theory and a (classical) truth-conditions theory need not be seen as rivals, if each restricts itself to its proper domain—cognitive role or reference, respectively.

The second general point to make is that the notion of semantic structure now falls into two. A key aim of a meaning-theory for a particular language is to explain and exhibit how the meaning of a sentence depends upon the meaning of its parts and their manner of combination.[10] If meaning has two components, then this project of exhibiting the structure of meaning will divide into two sub-projects: we shall want to know how the truth-conditions of a sentence depend upon the reference of its parts, and we shall want to know how the cognitive role of a sentence is determined from the cognitive role of its parts. (Frege can be interpreted as providing a sketch, for his notions of sense and reference, of how these theories of semantic structure might proceed.) Presumably the syntax required by each theory will coincide, but it is possible that the mechanisms of structural determination will be different in respect of the two components and thus call for different kinds of formal theory.

The third consequence relates to what might be called the topography of the total theory of meaning. Dummett is prone to picture a theory of sense (cognitive value) as surrounding, like a shell, the theory of reference—reference is depicted as derivative from sense (see especially Dummett 1976, p. 74). Put in my terms, this is tantamount to the idea that the component of meaning that subsists in the head determines and warrants the assignments made in the theory that deals with the extra-cranial component. Since Dummett's topography presupposes a wrong conception of how the components are related, it should I think be rejected as engendering an illusory sense of unity in the concept of meaning. A theory of cognitive role no more surrounds a theory of reference than a theory of desire surrounds a theory of belief in the ascription of reasons to an agent. (A question arises about the location of a theory of *force* on my view. I am inclined to hold that it properly attaches to the theory of reference not to the theory of cognitive role, since *what* is asserted (say) is that some state of affairs obtains in the world, and the speaker's representations are strictly immaterial to this—they are more expressed than asserted.[11] If this view of force is right, then Dummett is also wrong about the location of force with respect to the other two theories—again because he takes sense to determine reference. But I will not now digress to substantiate this position on force.) Putnam (to whose conception of meaning I am generally sympathetic) would not of course subscribe to Dummett's topography, but the gloss he puts on his own articulation of meaning as stereotype plus extension suggests a less radical position than seems to me indicated: he speaks of meaning as a *vector* of these two components, as if stereotype and extension somehow mingled in a unitary property (Putnam 1975b, p. 269). But I think that meaning is no more a vector of reference and cognitive role than $\langle 1, 2 \rangle$ is a vector of 1 and 2. It seems to me, then, least misleading to conceive a combined theory of meaning as simply the ordered pair of the two sub-theories, rather as a reason is aptly represented as an ordered pair of a belief and a desire. We shall then, in accordance with the recommended topography, construe each sub-theory as making distinct and independent assignments of reference and cognitive role to expressions, one theory allowing us to read off how the world is if the uttered sentence is true, the other supplying the materials for the (causal) explanation and prediction of a speaker's behaviour, specifically his use of sentences. A gnomic slogan capturing this conception of meaning might be this: the aim of each sub-theory is to specify the whole of part of meaning, not part of the whole of meaning.

I have so far characterized the two sub-theories in a schematic way; we need to know more about their precise form. The theory of reference will naturally take the form of a Tarskian theory of truth, possibly supplemented by some kind of account of the reference relation. This theory tells you, on the basis of the recursive structure of sentences, how the world is if the sentence in question is true—i.e. it gives *truth*-conditions, neither more nor less. In other words, a truth-theory is a specification of the *facts* stated by sentences of the object-language, in the intuitive sense of that recalcitrant notion. Given that this *is* the object of a theory of reference, what constraints should we impose on the truth-theory to ensure that it fulfils its appointed aim? It is notorious that material equivalence in a T-sentence is sufficient for its truth; but clearly this does not yield us the intuitive notion of fact. On the other hand, it would be wrong to require that the right hand side feature a sentence fully *synonymous* with the sentence mentioned on the left, since plainly 'Hesperus is a planet' and 'Phosphorus is a planet' have precisely the same conditions of truth despite their nonsynonymy. I suggest that an acceptable intermediate constraint is afforded by the condition of 'extensional isomorphism': the sentence mentioned and the sentence used in the theorems of the truth-theory should be composed of semantic primitives bearing the same extension in the same logical structure.[12] This condition is not without its complications, but I think it does approximate justice to the idea of the fact stated by a sentence; in particular, it is insensitive to differences of meaning between expressions traceable to cognitive role alone—how a sentence represents a fact should be kept distinct from the fact thus represented. The constraint can be put another way: if two sentences are intersubstitutable in all nonpsychological contexts—truth-functional, modal, explanatory, etc.—then the one sentence can be used to give the truth-conditions of the other. The thought behind this condition is that only psychological contexts are sensitive to the mental representations associated with the expressions in their scope (are representation-functional, as we might say) over and above extensional properties of expressions.[13] If these constraints do indeed serve to circumscribe the intuitive idea of fact, then they seem to me to offer a more hygienic explication of the fact idea than we have been taught to expect.[14] What should be emphasized is that these constraints on the truth-theory do *not* entitle us to replace the truth predicate and biconditional with 'means that' and preserve truth: but my whole point is that it is misguided to hanker after doing that, since a theory of truth-conditions cannot, by its very nature, deliver all there is to meaning—nor is this any deficiency in it construed as a theory of the whole of part of meaning.[15] If we want to capture the other part of meaning, we must move to a theory built upon qualitatively different principles—a theory of intra-individual cognitive role.

About this last matter it must be said that we have much less in the way of clear ideas and rigorously developed theory; perhaps, indeed, cognitive role is inherently insusceptible to the kind of formal theory exemplified by classical truth definitions. But I think that we have enough to appreciate what kind of thing cognitive role might be, and what a systematic semantic theory of it might look like. Out initial characterization of representations was in terms of their function in guiding behaviour in the light of evidence about the environment; so cognitive role will naturally be theorized in evidential terms. Field, in a paper (1977) to which I am much indebted, has built upon this initial (Quinean) thought a theory of cognitive role based upon the concept of subjective probability. Without recapitulating the details of Field's probabilistic semantics, his idea is that the cognitive role of a sentence determines and is determined by the speaker's subjective conditional probability-function on that sentence, i.e. his propensity to assign probability values to the given sentence conditionally upon other sentences. Exploiting this idea Field shows how to construct a formal semantics for propositional and quantificational languages, yielding soundness and completeness proofs. He also provides what is in effect a recursive definition of the subjective probability of sentences, showing how the probability of complex sentences depends upon the probability of their subsentences and subformulas. Sameness of cognitive role is defined as equipollence with respect to the speaker's conditional probability-function. According to Field, the cognitive role of an expression, as thus defined, depends upon something like associated descriptions (most clearly in the case of names), which may be improper or erroneous. The intuitive motivation for this is that what probability I assign to a sentence containing (say) 'Hesperus' conditionally upon the possession of certain evidence depends upon my *conception* of Hesperus—what I take to be true of it. Put in more Quinean terms, whether I assent to a sentence containing 'Hesperus' in response to impinging sensory stimuli depends upon the conception induced in me of the denoted object in the course of my acquisition of the name, i.e. upon what states of my head are associated with the name. Now, given this probabilistic account of cognitive role, we can ask how it fits in with our earlier observations about the independence of reference and internal representations: specifically, do the semantic assignments made in the probabilistic interpretation fix the assignments made in the truth-theoretic interpretation? We should predict not, since Field's associated descriptions can fail to determine reference in both of the ways discussed earlier; and indeed I think it does turn out that there is a clear sense in which a person's conditional probability-function on his sentences fails to fix their truth-conditions in the ways we have come to expect. First, as is most evident from the simplified Quinean theory, the conditional probability-function I assign to sentences containing 'water' will be the same as that assigned by my Twin-Earth counterpart to his sentences containing 'water', since we assent to such sentences in the same stimulus conditions (they are stimulus synonymous though referentially nonsynonymous). Similarly, indexicals used in evidentially indistinguishable situations will determine the same conditional probability-function, though the context may yield distinct truth-conditions. With respect to erroneous cognitive representations the failure of determination consists in a propensity to take sentences as evidence for the given sentence which do not in fact constitute evidence for the truth of the given sentence, i.e. which do not really raise the probability that its truth-conditions obtain. Thus suppose I associate 'the φ' with a name 'a', where a is not in truth the φ; then given a sentence 'the φ is F' I will be prepared to assent to (accord high subjective probability to) 'a is F': but of course it is not the case that the probability of 'a is F' is increased by the truth of 'the φ is F'. In other words, erroneous descriptions can cause my conditional probability-function to rate a sentence probably true in conditions that do not in fact afford evidence that the sentence *is* true. So here is a second way in which the truth-conditions semantics is independent of the probabilistic semantics. We thus have the desirable result that Field's theory of cognitive role has the consequence that the cognitive role of a sentence does not determine its truth-conditions. Nor is that at all accidental, since the two theories are constructed around quite different central concepts, the epistemic concept of subjective probability and the nonepistemic concept of truth. Relating all this to our conjunctive formulation of the dual contribution made by words in the 'means that' context, we can say in summary that the first conjunct is treated by a Tarskian truth-theory meeting the extensional isomorphism condition, while the second conjunct can be filled out by means of a subjective probability or evidential interpretation in the style of Field. Thus we have two central semantic concepts which supplement rather than rival one another; we have two theories of semantic structure, though based on the same syntax; and topographically the two theories exist *alongside* one another, not in any relation of inclusion. Furthermore, the cognitive role theory takes its place as part of a theory of behaviour, allied to decision theory; while the theory of reference does the job of recovering that aspect of communicative content which relates to conditions in the world. The former theory solipsistically characterizes the intra-individual role of representations; the latter nonsolipsistically articulates the relations between words and the world, ignoring how grasp of those words affects behaviour in virtue of their representational aspect.

Insisting upon the separateness of the two theories is not inconsistent with allowing that they might be combined or amalgamated to yield correct substituends in '*s* means that *p*'. To derive such meaning ascriptions we would first apply the truth-theory to *s* and then substitute expressions on the right hand side in accordance with the equivalences of cognitive role set up by the probabilistic semantics. Having thus taken account of both components we *can* replace 'is true iff' by 'means that'. For example, the truth theory might contain the theorem

'Hesperus is a planet' is true iff Phosphorus is a planet

this being a perfectly correct truth-condition; we then replace 'Phosphorus' by 'Hesperus' in the metalanguage on the basis of what the theory of cognitive role says about 'Hesperus'; we can then move to the direct meaning ascription. Two points should be noted about this method of amalgamation. First, we have got the effect of direct meaning specification without explicit use of the concept of meaning. Second, the extra conditions imposed by the theory of cognitive role do not make the truth-theory redundant in the enterprise of achieving full meaning ascriptions: this is because cognitive role does not determine reference and meaning does (think of giving the meaning of 'water is wet' in English and in Twin-Earth English). So there is a recipe for putting meaning together again; but what is the theoretical motive for doing so? Viewing the amalgamated theory as the object of the exercise encourages, it seems to me, an illusion of unity in the concept of meaning. Compare the two theories with a theory (if there could be such a thing) of what Dummett calls *tone*.[16] The tone of an expression is certainly a constituent of its conventional meaning, so that if two sentences differ in tone they do not substitute into '*s* means that *p*'. Tone is a separate and *sui generis* component of meaning, consisting perhaps in the affective connotations of an expression; it neither determines nor is determined by the other two components. A theory of tone would be a supplementary part of the total theory of linguistic significance, safely neglectable in developing the other two theories. To present a unitary theory, built around a single central concept, somehow combining all three elements, would obscure the true structure of meaning—suggesting as it would that each element could be theorized in the same terms. I think our attitude toward the relation between reference and cognitive role should be the same as our attitude toward the relation between tone and those other components of meaning. It would not actually be *wrong* to amalgamate the theories, but this should not blind us to their fundamental distinctness. I myself can see little theoretical point, and some danger, in presenting all of meaning within a single theoretical setting.[17]

The claim that meaning extends beyond states of the head is sometimes resisted on the ground that it implies that there are aspects of meaning which are not understood or grasped by someone with semantic competence in respect of the expressions concerned. That is, a nonsolipsistic conception of meaning denies that meaning is esssentially an object of semantic *knowledge*.[18] And it is felt that meaning could not be anything other than that which a person apprehends when he acquires mastery in a language; a nonsolipsistic view implies that the meaning of a man's words is opaque to him. (This is why it is supposed that reference could not be an *ingredient* of meaning.) This line of objection to our account can be reacted to in two ways. One way is just to concede that the psychological state of understanding a sentence does not fix its meaning—but then insist that this is unparadoxical once the two components of meaning are clearly distinguished. What *would* be paradoxical would be the idea that meaning is a unitary affair and yet only partially cognitive in character; but a difference in the epistemological properties of cognitive role and reference is in fact predictable from their essential nature. However, I prefer a rather different reaction. Understanding is itself a propositional attitude and as such can be expected to have content in the hybrid way in which propositional attitudes generally do: not all of understanding— sc. *what* is understood—is in the head. But then we *can* say that all of meaning is understood, since an extrinsic characterization of the state of understanding will encompass the extrinsic component of meaning. On this view of semantic knowledge you understand an expression just if (a) you associate the right representation with it, and (b) the resulting states of your head are appropriately related, causally or contextually, to the referent of the expression. So people on Earth and Twin-Earth understand 'water' differently after all. Understanding is comparable to perception in this respect. People on Earth and Twin Earth do not differ in the intrinsic aspects of their perceptual experiences, but their perceptual states are distinguished by being *of* different things. Understanding, we might say, is a *de re* propositional attitude. It is this that establishes a bridge of sorts between the theory of cognitive role and the theory of reference: understanding a sentence is a psychological state on which the assignments of the two theories converge— but each in their different ways.

V

In this section I shall briefly indicate the bearing of the view of content so far advocated upon some influential approaches to meaning; for the most part it will be fairly obvious what the bearing is.

(i)   *Pure truth conditions theories.* It follows directly from what has been said so far that any theory which attempts to explain meaning in terms of truth-conditions alone cannot be adequate; such a theory is constitutionally partial, because it omits internal representations. Thus possible worlds semantics (based upon *alethic* modality anyhow) will never add up to meaning, since it can register only one side of meaning.[19] This is because modal contexts are insensitive to the cognitive role of the expressions that occur in them. The lacuna shows up when possible worlds semantics tries to deal with psychological contexts and questions of cognitive value: its inability to account for these is not some peripheral defect, but results from the very nature of the theory—it has no place for the cognitive role of expressions. With Davidsonian theories the matter is less straightforward. Davidson proposes Tarski's Convention T condition of material adequacy upon definitions of truth as the proper object of a meaning-theory: the right hand side must *translate* the sentence mentioned on the left (see especially Davidson 1973a). There are many problems with this simple condition when we apply it to truth-theoretic semantics for natural language (e.g. indexical sentences), but the point on which I want to insist is that the requirement is too strong as a condition on definitions of *truth*, at least if translation is taken in its full sense, since it is no defect in a statement of truth-conditions that the expressions used differ merely in cognitive role (sense) from the expressions mentioned. It is indeed a defect in a theory of truth taken as a full theory of meaning, but what that shows is that we need to import a different set of considerations if we are to reach full meaning. The needed additional element is often introduced by imposing certain 'constraints' upon the core truth-theory in order to ensure the serviceability of the theorems in content ascriptions (Davidson 1973b; various papers in Evans and McDowell 1976). This way of looking at meaning is apt to mislead, however. First it should be clearly acknowledged that meaning cannot be reconstructed from truth alone, and the question then faced as to whether the needed extra element makes the truth-theory redundant. Second, we are given no *theory* of the needed extra element; the only theory in the offing is a theory of truth-conditions and this is agreed not to exhaust meaning. The proper verdict, I think, is that the Davidsonian perspective, while not being actually incorrect—for it is, after all, tacitly a dual component conception—is apt to deceive us about the theoretical resources we need in a fully adequate theory of meaning.[20] Certainly it cannot be glossed as claiming that meaning can be captured in terms of truth-conditions. One might almost say that the kind of bipartite theory suggested in the last section is the form the Davidsonian proposal would (or should) take if it were clearer about its own assumptions.

(ii)　*Pure use theories.* In recommending a use conception of content Dummett makes play with a revealing analogy. He compares the meaning of a word with the powers of a chess piece: meaning is exhausted in dispositions to use words, as the identity of a chess piece is exhausted by its role in the activity of playing chess (see 'The Philosophical Significance of Gödel's Theorem', p. 188, in Dummett 1978). Acceptance of this analogy can make the use conception seem incontrovertible, for we can indeed find no sense in the suggestion that the identity of a chess piece might transcend the powers it has on a chess board. Surely the game of chess could not be learned and the significance of a move appreciated if such transcendence obtained. However, the analogy is both tendentious and importantly misleading, for the simple reason that chess pieces, unlike words, do not have referential properties. Our conception of meaning is such that we regard Twin-Earth cases as possible, but we cannot coherently envisage Twin-Earth cases for chess pieces—their identity *is* supervenient upon the powers they possess. A word is more like a photograph than a chess piece in this respect: for there can be intrinsically indistinguishable photographs of different things. If we inserted such indiscernible photographs in some device whose output was sensitive only to intrinsic properties of them, then there would be no telling, on the basis of output, of which thing each was a photograph. This shift of analogy helps to dispel a related fear concerning the observability of meaning. Dummett argues that if meaning is to be communicable, it must be intersubjectively accessible; but if meaning transcended use, in the way the identity of a chess piece might (absurdly) be said to transcend its powers, then meaning would be something covert, occult and unobservable (see, e.g. ibid., p. 190). To deny that meaning is exhaustively manifest in use would be to make it into an ulterior trait of a speaker's mind, incapable of public exhibition. However, if the component of meaning that transcends use is reference, then the use-independent aspect of meaning would lie, not in the hidden recesses of the speaker's mind, but in public facts about how he is embedded in the world. Analogously, the intrinsic indiscernibility of two photographs need not make their representational properties undetectable: we can look to their causal ancestry to determine which objects are in fact photographically represented. Similarly, on a causal or contextual theory of reference, we could appeal to extrinsic facts about the speaker to recover the use-transcendent component of meaning. To deny that meaning supervenes on behavioural dispositions does not then have the dire consequences Dummett fears.

If the use conception is as flawed as I have claimed, then there is another sort of consequence for Dummett. This is that Dummett's case against a realist truth-conditions semantics, and in favour of anti-realism, is seriously undermined, since its crucial premise is precisely that meaning cannot transcend that which is manifest in a speaker's behavioural dispositions. It is significant here that Dummett's complaint against the causal theory of reference almost exactly parallels his objection to realist truth-conditions: that is, the notion of content delivered by both positions renders content incapable of behavioural manifestation.[21] Both views yield unacceptable accounts of understanding, because what is understood ceases to show up in dispositions to use. Consider a mental representation of a recognition-transcendent state of affairs; by hypothesis, what is thus represented will not be fully manifest in propensities to respond appropriately to evidence—the speaker's conditional probability-function will not exhaust the full semantic content of the sentences on which it is defined. So realist truth-conditions will not be explanatory of use. But from my perspective this is as it should be, since the use conception *can* only govern part of meaning. Let me put it

more strongly. To insist that meaning is use is to commit oneself
to methodological solipsism in semantics. Now if the content of
a sentence can consist in nothing other than what is explanatory
of use, then it seems that methodological solipsism in semantics
will lead to metaphysical solipsism. For the facts that explain use
are just states of the head—ultimately, dispositions to respond to
sensory evidence. So the content of a sentence will in the end
consist simply in the internal representations themselves, not in
the states of affairs represented—hence metaphysical solipsism.
What this whole line of reasoning ignores is that sentences have
referential properties independently of their role in the determination of behaviour. And without the basic assumption that
meaning is use, it is hard to see how Dummett's argument against
realism could be reconstructed. The general moral I would draw
from this is that, in assigning content to sentences and
propositional attitudes, we perforce rely upon our own theory of
the world in which we take the subject of assignment to be
situated; for part of assigning content is assigning reference and
we cannot hope to recover the reference of a speaker's words just
from truths about his behaviour. There is, in other words, no
neutral way to ascribe content—no way that is independent of the
theorist's own view of the world. (This point is relevant to the
question of alternative conceptual schemes: briefly, we have no
choice but to impose our own conceptual scheme on others—we
cannot bracket our own view of the world and try to discern
what the other thinks by putting ourselves into his head.[22])

Dummett and Quine have much in common; and Quine too
has taken to stating his behaviourist conception of language as
the doctrine that meaning is use.[23] More exactly, the meaning of
a sentence is identified with the set of sensory triggerings which
prompt the speaker's assent. Quine's view must then be that the
notion of reference has no role in a theory of meaning strictly so-
called; reference will not even be *determined* by meaning as he
construes it, since distinct referents can cause the same sensory
triggerings.[24] This leaves Quine with the question of what
theoretical purpose the relation of reference can serve in his
philosophy of language, in view of its redundancy in the
explanation of behaviour. I do not know that Quine has ever

seriously put this question to himself, nor what his answer would be. But it is at least clear that, on his view of meaning, reference is not any part of it.

(iii) *Translational semantics.* The conception of semantic theory put forward by J. J. Katz and his associates (e.g. Katz and Fodor 1963) is sometimes berated for failing to address itself to semantic relations between words and the world, this being the task of 'real semantics'.[25] I wish only to point out that this criticism is in a certain way misguided. For it is reasonable to construe translational semantics as some sort of attempt to give a theory of the cognitive role component of meaning, and a theory of that component *should* not be a theory of referential truth-conditions. Indeed, proponents of translational semantics would be equally justified in berating truth-conditions semantics for not doing the only thing that any semantics ought to do, namely characterize the internal cognitive role that words and sentences have. It is not that I think Katz's theory is a *good* theory of cognitive role; it is just that it is plausibly seen as engaged upon a perfectly legitimate enterprise, which is not to be criticized for not being the same as another (equally legitimate) enterprise. So the dispute between the two approaches seems to me to be based on a misunderstanding: there is no genuine conflict between them, because each is addressed to a different component of meaning; and neither can pretend to comprehensiveness.

(iv) *Gricean theories.* By a Gricean theory I mean any theory which attempts to reduce conventional sentence meaning to the propositional attitudes of speakers (cf. Schiffer 1973; Bennett 1976). A Gricean reduction will offer as an account of what it is for a sentence s to mean that p some collocation of attitudes at least one of which has the content that p. This immediately invites the suspicion that Gricean theories could not possibly be theories of what content consists in, since they presuppose the same content in their analysans as occurs in the analysandum; the Gricean cannot then be construed as in the same business as he who characterizes content in terms of cognitive role and truth-conditions. The theory might be offered, instead, as an account of the conditions under which public sounds and marks have the particular content they have—viz. when and only when they are produced with certain propositional attitudes. That interpretation of the Gricean programme—as telling what associates a content with a bit of language nonsemantically identified—seems to require that propositional attitudes themselves not consist in relations to sentences (internal or external) antecedently associated with semantic content, on pain of circularity.[26] To avoid the circularity the Gricean would have to claim that the representations involved in propositional attitudes are not properly linguistic, so that no sentence/meaning associations are presupposed. I cannot resolve the issue now: what I have wanted to point out is that the Gricean reductionist must confront the question of what it is for a propositional attitude to have content.[27]

A second issue concerns the conception of communication suggested by Gricean theories. At a superficial glance it might look as if the point of communication, for a Gricean as for Locke, is to convey to the audience the state of mind of the speaker as opposed to the condition of the world, since the speaker is said to intend the audience to believe that he (the speaker) has certain propositional attitudes. But in fact this implies the mind-centered view of communication only if the attitudes in question are themselves taken solipsistically. If, on the other hand, we view their content as implicating reference to things in the world, then it is not so clear that the Gricean has no room for the world-directed purport of acts of communication. We might say that the condition of the world gets conveyed *by* conveying the speaker's world-directed attitudes. What is perhaps more just is to accuse the Gricean of mislocating the *emphasis* in his implied picture of communication; he should acknowledge that the primary intention of a communicator is to let the audience know that the world is thus and so, the communicator's attitudes being an essentially secondary matter. My own view is that the content of an act of communication must be seen as comprising two elements, corresponding to the meaning of the sentence uttered: there is the information conveyed about the world, but there is also information about how the speaker represents the world, where this latter enables the audience to take what is communicated as usable in the explanation of the speaker's behaviour. The speaker's primary intention is indeed to discourse on the world, but he cannot do this except by revealing his own conception of it. Any adequate account of communication must make a place for both of these aspects.

(v)   *Kripke on names.* I would agree with Kripke (1972, 1979) that identity statements containing names are both necessary and *a posteriori*. But, given his own assumptions about the meaning of names, it is not easy to see how a statement *can* combine this modal status with that epistemic status. A natural account of the modal status of such an identity statement, favoured by Kripke, is that the names function simply to designate their bearers, so that the proposition expressed is aptly represented as an ordered triple consisting of the bearer(s) of the names and the identity relation. But this account of what is expressed by an identity statement leaves it problematic how the statement can have cognitive value, since the same representation of propositional content applies to statements of the form '$a = a$'. In short, Kripke's conception of the proposition expressed by an identity statement gets the modal status right but runs into trouble over the epistemic status. Suppose now that we start with the question of epistemic status and propose a classical Fregean account of the cognitively significant proposition (thought) expressed: the names are taken as semantically equivalent to a pair of proper definite descriptions, those the speaker or hearer associates with the names. Then the proposition thereby expressed seems to offer a plausible account of the epistemic status of the statement but (on reasonable assumptions about the sorts of descriptions speakers associate with names) encounters difficulties regarding its modal status, since the definite descriptions will (typically) not be rigid designators. The upshot appears to be this: neither the Kripke proposition nor the Frege proposition can do *both* jobs, yet both jobs need to be done. It can thus seem that an identity statement could not be both necessary and *a posteriori*. Kripke toys with the idea that the epistemic status might be explained by invoking descriptions only as reference-fixing not as meaning-giving, where such descriptions do not really contribute toward the proposition expressed. But I find this unsatisfactory for two reasons: (a) such reference-fixing descriptions are not—on Kripke's own showing—guaranteed to be available to the speaker, and (b) I think names like 'Hesperus' and 'Phosphorus' *do* differ in their meaning and not just in how their reference was fixed. It seems to me that Kripke's troubles here stem from his taking fixing the reference to be in a certain sense weaker than giving the meaning: he takes it that what fixes the reference of a name need not enter into its meaning, but that what does enter into meaning cannot but contribute to the determination of reference. This assumption has the consequence that if you try to account for epistemic status by reckoning descriptions into the meaning of names (perhaps idiolect by idiolect), then you will inevitably get the truth-conditions and hence the modal status of the identity statement wrong, since what constitutes meaning must also determine reference. The key to the problem is to give up that assumption; and this is exactly what the dual component theory does. We can then assign the epistemic status to the cognitive role component and the modal status to the referential component—and neither component determines the other. We could say that the identity statement expresses (independently) *both* the Frege proposition and the Kripke proposition; the problem arose because it was assumed that the statement has a unitary meaning.[28] Kripke makes the mirror image mistake to Dummett on this matter. Dummett is rightly anxious to account for the epistemic aspect of meaning, but is then driven to deny that the bearer of the name enters truth-conditions and so must resort to *ad hoc* stipulations regarding scope and related devices.[29] Kripke, on the other hand, wishes to account for modal status by letting the referents themselves occur in the proposition expressed, but is then hard put to handle questions of cognitive value. The common presupposition is that the role of descriptions in determining epistemic status must carry over to the determination of reference and truth-conditions. My suggestion is that, in rejecting this double role for descriptions, the dual component view of meaning can render a Fregean theory of informativeness *consistent* with a Kripkean conception of truth-conditions. Actually to accept the Fregean theory of cognitive value would, of course, involve repudiating the strict Millian view of the semantic content of names endorsed by Kripke; but I can see no advantage in retaining that view once it is clear that its rejection does not have the Fregean consequences for truth-conditions that Kripke fears.

The dual component view can also help us see what is going on in Kripke's puzzle about belief. The puzzle arises because we seem compelled to ascribe contradictory beliefs to someone in cases in which he expresses his beliefs using names acquired in different circumstances: in my terms, names with different cognitive role can invite the same ascription of belief-content. This is in fact a rather general phenomenon: it can also arise for demonstratives, as when a person assents to and dissents from 'that is F' at different times because he fails to realize that the same object is being demonstrated. I would offer the following diagnosis (which is not to say solution). First, because the singular terms in these cases have different cognitive roles for the believer it is easy to understand the state of mind in which such contradictory beliefs occur: the states of the head corresponding to the ascribed beliefs are not in fact the same. What would be very alarming and mysterious would be an ascription of contradictory beliefs in which the names on the basis of which the ascription is made are cognitively equivalent. This difference in the states thus ascribed comes out in their explanatory role, for the utterance of the sentences containing the terms will warrant quite different expectations about how the person will behave. So we can say, second, that in these cases the truth-conditions aspect of content dominates over the explanatory aspect: since the truth-conditions of the uttered sentences are precisely the same, we tend to ignore the cognitive difference between the names. Our usual principles of belief-attribution can, I agree, commit us to attributing formally contradictory beliefs, but this is not an inconsistency in internal representations—it obtains rather between the extrinsic truth-conditions. There is a sense in which the puzzle can arise precisely because belief-content has the two components we have distinguished.

VI

I have now motivated a dual component theory of content, sketched in the outlines of such a theory, and indicated its consequences for some standard views of meaning. It remains to take up a number of further issues for which the theory has implications. These are: (a) the ascription (or existence) conditions of content, (b) the nature of radical interpretation, and (c) the epistemology of content.

(a)   Under what conditions can we say of someone that he has a certain belief or of a sentence that it has a certain meaning? The dual component theory predicts that there are two sorts of condition, both of which are necessary and neither of which is sufficient: a belief-content may fail to be ascribable *either* because the person is wanting in adequate representations *or* because the world does not contain such objects and properties as would confer truth-conditions on the belief; and similarly for the psychological state of knowing meaning. So it might be thought a test of the rightness of the view I am advocating whether we do indeed refuse to assign a content in cases in which either the cognitive role or the reference is lacking. We have then two sorts of case to consider: those in which a person's assent to a sentence is not backed up by an association of adequate representations but reference is present, and those in which assent is accompanied by representations but the sentence lacks referential truth-conditions. On the view that words in content clauses make a dual contribution it should not be possible to ascribe content in the usual way in either sort of case. Cases of the first type are of two sorts: those in which no representation is correlated with some word(s) in the assented-to sentence, and those in which some correlated representation is wildly in error. The former type of situation can be generated by (broadly speaking) the division of linguistic labour: a person merely mouths a sentence to which others in his speech community attach adequate cognitive representations (stereotype)—an extreme case would be passing on a coded message undeciphered by the person transmitting it. Here the words can have reference—even in the mouth of the person in question—but no belief expressed by the sentence is ascribable to the person; his conception of the sentence's subject matter is too impoverished. This lack of associated representations makes itself felt when we try to use the assent to explain the person's behaviour by way of a belief-ascription: his psychological state will not match that of someone whose behaviour can be explained in terms of the belief in question. The latter sort of case arises when a person severely misunderstands the sentence to which he assents. Clearly someone who thinks that 'bachelor' applies to married females cannot be ascribed a belief whose content is specified by using that word. I am inclined to be liberal about how much error it takes to obstruct a content ascription, but there are limits beyond which no belief can be ascribed. In these cases the person's assent is again psychologically idle—he will lack the usual dispositions in respect of evidence and action. I think the absence of explanatory potential is the reason we are unprepared to ascribe beliefs in these cases, though the sentence itself is possessed of truth-conditions. But now are adequate representations *sufficient* for a content ascription? Someone (like Frege) who thinks that cognitive role determines reference will suppose so, since the referential component will be recoverable from the representations; but since I deny such determination it looks as if I *have* to say that empty reference blocks content ascription. For an ascription of belief involves, as we said, a claim about the (quasi-) semantic relations between the subject's representations and things in the world; if that claim is false it follows that the subject has no belief with that content (compare factives). So if a sentence does not have referential truth-conditions it cannot be used to ascribe a belief-content. We thus have the consequence that the nonexplanatory aspect of belief is also necessary for the existence of content. Taking both sorts of case together we get the predicted result.[30]

That content existence depends upon both components helps explain and reconcile what have seemed like contrary intuitions. Some have claimed (rightly) that since content is determined by the extra-cranial world, beliefs will not be available unless the world is a certain way; it follows that beliefs have no purely intrinsic characterization (McDowell 1977, section 8). Others have felt that this flies in the face of an obvious truth: from the point of view of the explanatory role of a belief, a person's dispositions to behaviour are invariant with respect to how the world is—so beliefs *must* be in the head (Stich 1978b). The disagreement results from each side concentrating on one aspect of belief to the neglect of the other. I think it is correct to insist that the relation between subject and object is necessarily mediated by representations whose existence is world-indifferent—for an object cannot be its own cognitive role—but it is wrong to infer that belief-content can be solipsistically ascribed. If God could look into (the intra-cranial part of) our minds he would indeed not see there what we are thinking of: but that is not because he would see *nothing*—he would see our internal representations—but because what is there does not suffice to determine the objects of our thought.[31] In other words, beliefs do have an intrinsic component—they must have if they are to be causally explanatory—but that component cannot be identified with the belief of which it is a component, since content also requires reference. The existence conditions of content are like those of an ordered pair: the pair does not exist unless both elements do, but it does not follow that the elements can be *identified*.

(b) What is the empirical status of a theory of meaning for a particular language? Since I have divided the total theory into two sub-theories, the question of radical interpretation now comes in two parts, corresponding to the assignment of truth-conditions and the assignment of cognitive role. It is obviously an interesting question then whether the two sub-theories are independently testable. Before I address this question let me make some remarks about its significance. It seems to me that the question of empirical status is secondary to the question what *form* a theory of meaning ought to take: given a certain conception of meaning, the proper view of radical interpretation should fall out as a consequence—we should not let the verification conditions of a theory of meaning dictate its internal conceptual structure. To do so would be to embrace a positivism that is no more reputable in the theory of meaning than elsewhere.[32] What is more significant for theorizing meaning is the question of the *point* of the notion of meaning—or of the components that make it up—for this question will better lead us to the concepts in terms of which meaning is to be theorized. If it should turn out, then, that the two sub-theories could not be independently tested, we ought not to conclude that our theoretical conception of meaning is somehow defective—that it alleges a theoretical difference where there is no empirical distinction—rather, we should doubt the significance of radical interpretation in the theory of meaning. (We would not feel tempted to revise our customary articulation of reasons into desires and beliefs just because the attribution of these attitudes turns out to be empirically interdependent.) However, that said, it does seem reasonable to expect that the already noted differences between reference and cognitive role will be reflected in diverse modes of empirical verification. Let me first indicate, by way of a foil, how (idealized) radical interpretation would proceed under two other dual component theories—Frege's (as seen through Dummett's lens) and Putnam's (in 'Reference and Understanding' (1978)). On Frege's view of sense and reference, the two theories would be fundamentally interdependent in their conditions of confirmation. For once we have arrived at a theory of sense—construed as that in which cognitive significance consists and as what explains our judgements—we will already have settled upon a scheme of reference for the language, since sense determines reference. For the same reason it is *necessary* in order to arrive at a reference assignment, that the interpreter already have a theory of sense; there will be no independent way of getting at reference. By contrast, Putnam's theory of total meaning would render the two theories fundamentally independent in point of empirical confirmation. For the theory of use and understanding can be tested without any thought of what accounts for the success of language behaviour, and vice versa—just as describing the competence involved in turning on lights is independent of a theory of electricity. It seems to me that Frege's view makes the empirical status of the two theories too interdependent, while Putnam's picture suggests an unrealistically radical separation. I can indicate my own view of the matter by considering perception. Here we can envisage a pair of assignments to a perceiver made over time: first, an assignment of experience characterized in 'as of' terms; second, a relational assignment of objects perceived, this done in the transparent style. It is clear enough that knowledge of one assignment would not *yield* knowledge of the other: but could we come to have knowledge of the one without any knowledge of the other? In practice, no doubt, we rely upon information about the actual properties of surrounding objects to conjecture how someone else is perceptually representing those objects—what he perceives them as—and we also take reports of perceptual experiences to be (typically) good indicators of the actual condition of external objects of perception. But is it possible *in principle* to verify such perceptual assignments independently? It seems to me that it ought to be. We could test our assignment of experiences on the basis of how, in combination with other mental states, they would (causally) influence behaviour (we might in this way assign a course of experience to a man we know to be constantly hallucinating). On the other hand, we could judge whether an object is (transparently) seen on the basis of (to put it crudely) how the perceiver is spatio-temporally embedded in the world and in what causal transactions he is involved. In something like this way we can envisage separately testing a theory of reference and a theory of cognitive role. In respect of the latter, we will, in the manner of Quine, observe how our subject responds to sensory stimuli, thus establishing his conditional probability-function—and all this within narrow psychology. This procedure will not yet have issued in the verification of any assignment of reference, for the reasons already given. How then is the reference theory to be tested? Suppose we have a theory of *reliability* for speakers, i.e. a theory of the conditions under which speakers learn (acquire true beliefs).[33] Conjoining this reliability theory with a candidate assignment of truth-conditions the empirical content of the conjunctive theory will be this: the theory is empirically warranted if the world turns out to be as the theory of reliability and truth-conditions jointly predicts. Let the conjunctive theory say that a sentence s is true iff p and that the belief expressed by an utterance of s was reliably acquired; then if it is in fact the case that p the theory's assignment of truth-conditions is (to some degree) confirmed—otherwise it is disconfirmed.[34] Whether a reference theory is totally empirically independent of a cognitive role theory clearly depends upon whether we could have a theory of reliability that made no assumptions about cognitive role. It is difficult to resolve this question without a better idea of the precise character of a theory of reliability, but it does at least seem evident that *some* reliability considerations can be ascertained without enquiry into a person's inner psychology—we can tell, e.g., whether external circumstances are conducive to unimpeded observation. (The same conclusion appears indicated by so-called causal theories of reference.) With respect to propositional attitudes, verifying the two components of their content can be expected to combine the methods appropriate to perception and to meaning, in view of the location of belief *vis-à-vis* those other two sorts of property. However, I do not want to be dogmatic about this: it *may* be impossible properly to know the referents of a man's words without knowing how he represents those referents, and vice versa—so that (as with joint attributions of beliefs and desires) we need, at some stage, a method of simultaneous determination. But at present I can see no very convincing argument for this and some reason to doubt it; certainly the two sub-theories can be supposed, in large measure, to have their empirical consequences in isolation.

(c)   Dummett says: "It is an undeniable feature of the notion of meaning—obscure as that notion is—that meaning is *transparent* in the sense that, if someone attaches a meaning to each of two words, he must know whether those meanings are the same.' (1978, p. 131). It will be evident from what has been said up to now that I am committed to denying what Dummett takes to be undeniable. But why should anyone hold that meaning is thus diaphanous? (I prefer this term, as 'transparent' already has a technical usage.) On the face of it the claim seems questionable in view of its apparent conflict with a second undeniable feature of meaning, namely that meaning determines reference; for no one could hold that *reference* is diaphanous. Meaning appears constitutionally diaphanous because of the familiar assumption that what is in the head determines reference, i.e. the reference relation is internally represented. But once we reject, as we have, the assumption that cognitive value is the mechanism of reference, the second undeniable feature of meaning immediately controverts the first: words can have the same cognitive role yet differ in reference and hence in meaning.[35] Since reference is not supervenient on cognitive role, an identity statement can be potentially informative even though the terms in it are cognitively equivalent. (Difference of cognitive role is not then necessary for informativeness, though it does seem sufficient.) Denying that total meaning is diaphanous does not, of course, imply that there is *no* dimension of meaning which has that epistemological property; and indeed it seems very plausible that cognitive role *is* diaphanous—for it resides in the head and can therefore become accessible to introspection. Since it is not the case that all of meaning is diaphanous it is not a consequence of incorporating reference into meaning that one who understands an identity statement must know its truth-value— understanding relates to reference in an extrinsic way. The semantic content of a sentence is analogous in this respect to the representative content of a perceptual experience: for what is perceived includes the external object of perception, and so is not something wholly accessible to introspection. Acknowledging the duplex structure of content in general thus prepares us to reject the diaphanity thesis.

We can now detect some deep connections between a number of mutually supporting doctrines about meaning. Methodological solipsism in semantics, meanings as introspectible, the use conception of meaning, description theories of the reference relation—all of these reflect an underlying *individualism*: content must be supervenient upon properties of the individual (inner or behavioural) taken in isolation from his environment (cf. Burge 1979a). Thus if meanings were in the head, we could expect them to be both diaphanous and exhaustively manifested in use; description theories are then offered to preserve these doctrines from the threat posed by the semantic relevance of reference and truth. As soon as we take a different view of the mechanism of reference these behaviourist and introspectionist conceptions of content come to seem mistaken. I suspect that attachment to these doctrines arises out of a tacitly Cartesian approach to content: what we say and believe must be accessible from a first person point of view. The alternative position here advocated tends to approach the matter from a third person viewpoint: the individual is seen as situated in the world in such a way that the semantic content of his actions and psychological states results jointly from his intrinsic properties and his relations to the world. Content is conferred as much by the world as by us; the epistemology of meaning reflects this fact.[36]

Notes to chapter 5

[1] So dubbed by Putnam (1975a). A similar idea lies behind Saul Kripke's talk of epistemic counterparts in 'Naming and Necessity' (1972). See also, for an interesting extension of the Twin-Earth idea to the social environment, Burge (1979a).

[2] Actually, Fodor (1980) doubts whether the latter type of psychology is feasible. His reason seems to be that we would need to have a more or less complete scientific theory of the nonpsychological world if we were to characterize referents in ways which are (a) nomological and (b) uniquely identifying. This reason appears weak to me; the problem is that it proves too much. For the same considerations would rule out a science of photographs, or of ecology. (Not that I want to say that we could have a genuine *science* of propositional attitudes.)

[3] This clearly bears upon the conception of rationalization advocated by, e.g., Davidson (1963a). Indeed, it now appears that the causal factors implicated in acting on reasons are distinct from the truth-evaluable propositions that determine the logical relations between propositional attitudes. So even if the whole of content *could* be shown to have explanatory relevance (which I have denied), we would still have to qualify the claim that rationalization is *causal* explanation. I would say that not all of rendering an action 'intelligible' by ascription of reasons can be assimilated to specifying the conditions that causally explain it—though a component can.

[4] This consequence is most explicitly acknowledged by Putnam in 'Reference and Understanding' (1978).

[5] As Putnam (1975a) pointed out, it follows that meaning is not in the head, on the assumption that meaning determines reference.

[6] Dummett (1973), p. 147. We might say that, for Dummett, the reference relation has semantic relevance only if it is dispensable in favour of nonrelational conditions of the speaker, i.e. is not really a *relation* between word and object.

[7] I cannot cite a published source for this thesis, but it is prominent in the (recent) oral tradition.

[8] I suppose the *locus classicus* of the view is Book III of Locke's *Essay Concerning Human Understanding*.

[9] This connects with Jonathan Bennett's claim (1976, pp. 84f), that a necessary condition of belief-ascription is educability: if I am right, educability is required for aboutness.

[10] This was Davidson's official aim in 'Truth and Meaning' (1967). In that paper, however, Davidson is (rightly, I think) cautious about the claims of a recursive

specification of truth-conditions to be a complete theory of (intuitive) meaning; so strictly, all he can claim is that a Tarskian truth theory shows how *one component* of the meaning of a sentence is structurally determined, viz. the referential component.

[11] Cf. a remark of Schiffer's (1978, p. 176). This paper gives some cogent reasons for interpolating modes of presentation between subject and object, but I do not agree that these determine (and so make redundant) mention of the object in giving the content of a belief or sentence.

[12] This is, of course, the extensional analogue of Carnap's relation of intensional isomorphism: see Carnap (1956), section 14.

[13] I thus commit myself (gladly) to a divided account of the truth-conditions of (alethic) modal contexts and psychological contexts. The division is predictable from the 'Shakespearean' character of the former in contrast to the latter: i.e. co-denoting names (but not definite descriptions) are intersubstitutable *salva veritate* in modal but not psychological contexts.

[14] E.g. by Davidson (1969). In saying this I do *not* intend to resurrect the idea that facts are a peculiar kind of *entity* quasi-designated by whole sentences; I wish only to point out the semantic transparency of ascriptions of truth-conditions and note its significance.

[15] My point here can be put this way: it is a mistake to think that if the notions of truth and reference cannot be made to deliver *all* of meaning, the proper response is to *start again* with some other set of semantic notions.

[16] See Dummett (1973), ch. 1. I recommend reflection on tone as a prophylaxis for those inclined to hold, as a dogma, that meaning *must* be a unitary matter.

[17] I am thus opposing, as potentially misleading, the Davidsonian conception of the form of a complete theory of meaning, most clearly exemplified in McDowell (1977)—more on this later in the present paper.

[18] See Dummett 1973, p. 92, and 'The Social Character of Meaning', in Dummett 1978.

[19] David Lewis wrestles (unsuccessfully) with this problem in 'General Semantics' (1972a, pp. 182f). The notion of proposition generated by this approach to meaning—as equivalent to a set of possible worlds—clearly unsuits propositions so defined to serve as the internal representations needed to account for the causal–explanatory aspect of belief.

[20] There is a danger, too, of a certain kind of circularity in appealing to constraints from propositional attitudes with content: for what we wanted was a radical theory of what content comprises. (Cf. my remarks on Gricean theories in section V(iv) of the present paper.)

[21] On realist truth conditions and use see 'The Philosophical Basis of Intuitionistic Logic' in Dummett (1978), and see also Dummett (1976). On the causal theory of reference and use see Dummett (1973), pp. 146f. These two questions seem to me to be intimately connected.

[22] This puts me in agreement with the conclusion of Davidson (1974a), but I reach that conclusion by a diametrically opposed route.

[23] Especially in his 'Use and Its Place in Meaning', in Margalit (1979). It is also worth noting, apropos of my earlier interpretation of Dummett's use conception, that he (Dummett) expresses 'strong agreement' with Quine's insistence on reducing knowledge of meaning to dispositions to behaviour: see Dummett (1979), p. 134.

[24] See Quine (1960), p. 31, where in effect this is acknowledged. There may, indeed, be some irony in this for Quine: meaning may be a more infirm notion than reference in point of clarity, but it appears to have a surer theoretical role in Quine's philosophy of language than does reference—it being what accounts for behaviour.

[25] Thus David Lewis (1972a, p. 169). (I do not, however, dispute the point that knowledge of a translation manual is insufficient for knowledge of meaning.)

[26] Field raises this difficulty for the Gricean (1978, pp. 52–3), but I think underestimates its potential force: he takes it that circularity is avoided if (but only if) the semantics of the internal language can be developed independently of the semantics of the public spoken language. But even if that were possible (which he doubts), the fundamental aim of the Gricean project would be frustrated, the aim being to demonstrate that semantics can be (analytically) reduced to psychology—to the contrary, psychology would rest upon semantics.

[27] This question is especially pressing for those who wish to explain linguistic truth and reference in terms of propositional attitudes: e.g. Christopher Peacocke (1976).

[28] Indeed the very idea of proposition comes to seem to embody a mistake from the present perspective: for there is no *one* thing that is both the bearer of truth-value and that which determines the cognitive import of a sentence, i.e. its psychologically explanatory aspect.

[29] See Dummett (1973), ch. 5, Appendix. Christopher Peacocke (1975) seems to me to make the Kripkean mistake.

[30] Here is perhaps the place to insert a remark about functionalist views of propositional attitudes; as, e.g., in Schiffer (1978). The intuitive idea of functionalism is that a mental state is individuated by its causal role in the agent's psychology. On my view of content, a component of it eludes capture in these terms—for people can have attitudes possessed of different contents and yet be functionally isomorphic. A possible response to this, suggested by Field (1978), is to try for a functionalist account of the reference relations constitutive of that component of content that goes beyond individualistic functional properties. However, I think this reponse rescinds the intuitive motivation for functionalism: we would have to think of the whole world, not just the individual, as a functional system. Cf. McGinn (1980), section V(iii).

[31] McDowell appears to hold that nothing of the mind is discernible within the skull (a Sartrean view). But admitting that the truth-conditions of a belief introduce extra-cranial conditions does not require us to accept the extreme view that the mind is not in the head. Indeed that view cannot, I think, properly explain how propositional attitudes are mental causes of behaviour. My position is that *some* of the mind is located in the head. (This way of putting it is due to McDowell.)

[32] The prime target of this remark is Quine, but I think Davidson falls into this way of thinking sometimes: see especially Davidson (1977).

[33] Such a theory is alluded to in Putnam (1978) and Field (1978). A reliability theory plays essentially the role sometimes allotted to the principle of charity, e.g. by Davidson (1973b). That principle would also provide a rule for assigning referents—viz. maximize true assertions and beliefs—that will not immediately yield information about the speaker's representations. However, I think the principle is unacceptable: see McGinn (1977).

[34] That is, the interpreter tests for whether he could acquire knowledge of the *world* by means of the candidate theory of referential truth-conditions; note that this method has the merit of conforming with the *point* of communicative speech acts.

[35] I hold this not only for the inter-speaker case and the cross-temporal intra-speaker case, but also for the individual speaker at a given time. Surprisingly, Field (1977) stops short at this last case: he says (on p. 396) that the reference of an individual's words at a given time should be determined by the states of the individual's head at that time, and he proposes a constraint to ensure this. However, it seems clear to me that the usual ways of constructing counterexamples to this determination thesis for the other cases—with

respect to names, natural kind predicates and indexicals—are equally available in the single-time intra-speaker case; I leave the exercise of constructing such examples to the reader.

[36] My debts in this paper will be evident from the notes, and the sources mentioned in the text, but I must acknowledge a special indebtedness to the ideas of Hartry Field; Anita Avramides, Jim Hopkins and Christopher Peacocke made helpful comments, as did auditors of talks I gave at the Universities of Birmingham and Cambridge. I should say that I intend this paper as an interim study, rather than a terminally definitive piece. (Those readers who sense a tension between the claims made here and some suggestions made in other papers of mine are not wrong.)

6

On Specifying the Contents of Thoughts

Andrew Woodfield

Can one person specify exactly what another person is thinking? An affirmative answer does not seem totally out of the question. But on reflection, the notion of a 'thing thought' begins to seem problematic. We have some inclination to believe that psychological states can be measured and described accurately. If intentional states are *real*, as they surely are, they must be studiable scientifically. But the *intentional content* of a mental state is not like duration, intensity or any other magnitude familiar in science. It is in the same sort of category as a meaning or a gist. Perhaps specifying the content of a thought is like conveying the gist of a poem, where the notions of perfect accuracy and completeness do not readily apply.

This paper tries to explain what speakers are really doing when they specify the content of some subject's thought. The key idea is that the would-be specifier, while using his or her language as an instrument for communicating his or her own thought, also uses a language as a *model* of the mind of the subject. The language functioning as a model is usually the same language as that in which the communication is couched. But the languages could be different, and the two kinds of use are more clearly distinguishable when this is so.

It is not the aim of the paper to propose a theory of what it really *is* for a certain thought to have a certain content. But it is inevitable that an account of what it is to specify content should be backed up by some view about what content is, and indeed I do make one or two assumptions on this score. For example, I assume that *thinking that p* is, at least in part, a matter of being in an internal state which has a characteristic role in reasoning. But it is not claimed that functionalism tells the *whole* story about content, as far as conscious thoughts are concerned.

To attribute content to a psychological state is to treat it as a representation. Representations can be classified in two ways; by what they are like (the nature of the *representans*), or according to what, if anything, they point to (the *representandum* or *representanda*). The two methods are different, so there can be no homogeneous science of 'representationology'. This paper is concerned with the former kind of classification. I take the content of a mental state or episode to be a property of a *representans*. It belongs to the state or episode solely in virtue of the way the subject S is psychologically structured. Thus the theory of content, as here understood, conforms to the principle of *methodological solipsism* (Putnam 1975a; Fodor 1980). Our investigation hinges upon the way *inner contents* are to be specified.

The present approach eschews any attempt to specify contents by reference to *representanda*, be they Meinongian objectives, Russellian complexes, possible facts, Bergmannian states-of-affairs which may or may not be actualized, or propositions construed as entities in a world (see Aquila 1977 for a definitive critique). Such theories embody what by my lights is a confusion between the intentional content and the *object* (construing 'object' liberally here). The content would be whatever it was even if there were no object. Our view stands in direct opposition to the view that intentionality is essentially the property of being directed upon an object (cf. Searle 1979a; see also Woodfield 1976, especially pp. 204–5).

The approach also differs from, but is perhaps complementary to, that which seeks to specify the content of a cognitive state via the specification of conditions that the world would need to meet if the state were to represent reality truthfully. Such a method is indirect. It tries to pin down content by specifying a way the world must be when the content matches it, instead of specifying what the representans itself is like (cf. Dennett's 'notional worlds' (this vol.)). The present approach likens intentional contents to intensions viewed as logico-semantic roles of terms and sentences within a language-system, not to intensions viewed as determiners of extension, reference or truth-value. This distinction corresponds to the distinction between conceptual role (functional) semantics and truth-conditional (referential) semantics (Field

1977; Loar 1976b, 1980; McGinn, this vol.). A full defence of the 'functional semantic' strategy will not be attempted here, but I suspect that it is impossible to develop a theory of contents *qua* fixers of truth-conditions, because there is no feature contained in the cognitive state which determines a unique truth-condition for that state. However, this is a tricky issue, which revolves partly around the question of what exactly is meant by 'truth-condition'. I prefer to try the 'functional semantic' approach because it facilitates a view of representational psychology as having two conceptually distinct parts: one part is the theory of what is inside the subject's mind, how its contents divide up and how they interact; the other is the theory of the relations between the subject and his environment (including causal transactions over time). This distinction was drawn, and explored to great effect, by Hartry Field (1977, 1978).

Nevertheless, a huge question-mark hangs over the theory of content. It is not that intensions *qua* roles are particularly suspect *ontologically*. They are no more suspect than *routines* or *procedures*, for example. The big trouble is that the term 'content' is *metaphorical* (as Frege pointed out).[1] The metaphor has first to be interpreted properly, then dispensed with, otherwise the study of states individuated by content is doomed to remain pre-scientific.

For convenience I shall assume for the most part that the things up for classification into content-types are datable acts or episodes of *thinking, judging* and *apprehending*, acts of the sort that might be ascribed to a subject by means of sentences of the form 'S thinks (at time t) that p'. I shall call such acts and episodes 'thoughts', adopting an obviously non-Fregean usage. In Frege's terms, a 'Thought' was more like what I call the *content* of a thought, except that Fregean Thoughts are essentially bearers of truth-values, which makes them unsuitable for purely psychological purposes (cf. Stich 1978b; Perry 1977; Burge 1979b; Fodor 1980). Thinking-episodes are useful as *classificanda* because ordinary discourse treats them as individuals. Also one avoids getting into arguments about 'implicit' or 'virtual' beliefs (cf. Dennett 1975) for, presumably, occurrent thoughts are, as A.I. people put it, 'explicitly represented'. Still, the problem of specifying content arises for all the so-called 'propositional attitudes', so the conclusions to be reached should be generalizable.

The *dramatis personae*, in addition to the subject S, will be an ascriber A of thoughts to S, plus a hearer H of such ascriptions, who share an *ascription-context*. Their context may or may not be identical with the context in which S does the thinking. S's context consists of S's real-world environment, plus the psychological background of S's previous and concurrent beliefs, desires, feelings, sensations etc.

I. Preliminary remarks about specifications in *oratio obliqua*

When people classify thoughts into types, the criteria vary greatly according to the purposes for which the classification is made. Suppose two contestants are being quizzed on their knowledge of capital cities around the world. One contestant has the thought that Buenos Aires is the capital of Argentina, the other thinks that Brasilia is the capital of Brazil. The two thoughts are similar in certain respects and different in certain respects. For example, someone interested in a certain similarity could classify them both as being of the type: *thought about a Latin-American city to the effect that it is the capital of a certain Latin-American country.* Someone interested in a difference could point out that the former is of the type: *thought about Buenos Aires*, the latter of the type: *thought about Brasilia.*

Another method is to turn the complete 'that'-clauses written above into names of types. Thus the first was of the type: *thought that Buenos Aires is the capital of Argentina*, the second was of the type: *thought that Brasilia is the capital of Brazil.* The earlier methods were obviously derivative from this, in that they exploited the meanings of the words in the original 'that'-clauses.

Only a tiny proportion of the thoughts had by a person are actually ascribed. Nevertheless, the unascribed ones could in principle have been ascribed, and the ascriptions could have been in *oratio obliqua.* Hence unascribed thoughts belong to types nameable with the help of 'that'-clauses, even if the 'that'-clauses are never formulated by anyone. As soon as a classifier turns his attention to such a thought and assigns a 'that'-clause to it, his act of classification *is* an act of ascribing. So the would-be classifier or specifier just *is* someone occupying the A-slot who has taxonomic intentions.

However, every thought belongs potentially to a host of distinct 'that'-clause types. For example, it would be acceptable, in appropriate conversational contexts, to say any of the following: (a) The second contestant thinks that Brazil's capital is Brasilia; (b) She thinks that Brasilia is the capital of a certain country; (c) She thinks that Brasilia is the capital of the largest country in South America; and so on. Not all of these specify the inner content of the thought. Perhaps none of them does. Once again, the variety of ascriptions reflects the fact that ascribers may have different interests and purposes. Sometimes they are interested only in the object S thought *about*, sometimes in the object as described in a certain way, sometimes in a *part* of the content, sometimes in the object *plus* part of the content, and so on. What the classifier of pure contents needs is a way to weed out the hybrids and isolate those ascriptions which specify the whole content and nothing but the content *as it presented itself to S.*

Could there be an ascription which did not reflect the interests and purposes of the ascriber? The question is ambiguous. In one sense, every act of ascribing 'reflects' a purpose in so far as it is performed for a purpose. There must always be a reason for A's choosing one form of words over another. The would-be specifier of content is no exception. He believes that each thought has a constitutive inner content and his purpose is to group thoughts into basic types on that basis. The taxonomy he wants is standardly Aristotelian. No thought will belong to more than one *basic* type or *infima species*, each belongs to its basic type essentially, and no two thoughts that differ in any contentful respect belong to the same basic type, though there may be higher-level groupings to which they both belong, just as your dog and my cat belong to the kind 'mammal', and also to the non-natural kind 'pet'. His purpose is to classify in a way that captures real distinctions and similarities among the *classificanda*.

In the other sense of 'reflects', a typology reflects the classifier if some aspect of him or his activity enters into the conditions of identity of the types that he postulates. Construed in that way, the question is asking whether A, in ascribing, is endowing S's thought with a character it did not have before. If so, ascription is never specification of a feature that existed independently. It is the creation of something new. The term 'specifying' is simply a misnomer for this kind of innovative hermeneutical activity. We would be forced to conclude that no one in the A-role can botanize in the objective, naturalistic way envisaged by our would-be specifier.

Such scepticism is unacceptable, however. The practice of interpersonal ascription of mental states is built on certain presuppositions which are incompatible with that view. One such presupposition is that the content of someone's thought on a given occasion was what it was, irrespective of whether any other person ascribes it, thinks about it or knows of its existence. Another is that if A tries to discover what S thought, A's efforts and opinions cannot make it the case that S thought one thing rather than another. Ascriptions are hypotheses to be tested against evidence of various kinds in accordance with standard procedures, just like any other empirical hypotheses. A competent ascriber knows all this and allows himself to be guided by it. There is no reason for doubting that S's thought-contents are ontologically independent of A and A's views (where A ≠ S), so it is a perfectly coherent enterprise for A to try to specify them.

But A's words have to *function* in the right way. A remarkable feature of *oratio obliqua* is that the words in A's report can perform at least three separate functions. On any given occasion of utterance, a word or phrase might be performing just one, or two, or all three together. To illustrate: A, while talking to H amongst a small group of people which includes Sally, says 'Sally is thinking that the coloured man is Brazilian'. Let us focus on the definite description. First, it could be performing a referential or back-referential function (these are distinct, but it is convenient to group them together). If A's utterance is not the continuation of a previous conversation, and if the circumstances are normal, A is using the description to pick out someone in the group for H's benefit. If the utterance is part of an earlier conversation, A may be harking back to a topic introduced by an earlier term. The earlier term may have occurred in *oratio obliqua* or *oratio recta*. It may refer to an actual person, but need not—A and H may have been discussing the characters that Sally plans to include in her next novel. What the referential and back-referential uses have in common is that they aspire to pick out a topic in the context of discourse shared by A and H. By putting the description inside the 'that'-clause, A signals that the topic referred to is also a topic of Sally's thought, but A is not claiming that Sally is in a position to exploit the description in the way he intends H to exploit it, that is, to identify the topic. For Sally may lack some of the supplementary background knowledge that A invokes in H.

Secondly, A may be telling H that Sally is thinking of someone *as* a coloured man. In this capacity the description serves to specify part of the content of Sally's thought, and it does this in virtue of having a *sense*. H has to understand its sense to gain access to a concept which was active in S's mind at t (according to A's hypothesis). (Of course, in cases where the description has a referential function, its sense helps it to perform that function too.) The description performs the content-specifying function if and only if it occurs *opaquely*, in the sense that it cannot be replaced, without risk of changing the truth-value of the whole sentence, by any coreferring term having a different sense.

A might utter the description not intending it to be specificatory of part of S's thought-content. People often speculate about a subject's unexpressed thoughts without venturing to specify their inner nature. They lack sufficient evidence of what went on in the subject's mind. So H may have every reason to construe what A says just referentially or back-referentially. Still, what A said might, in fact, be a true specification of content, even though neither A nor H knows this. What counts is the sentence A uttered rather than A's intention in uttering it. The question whether a given sentence qualifies as a specification is actually *independent* of whether any term in it is functioning in the first way on the particular occasion of utterance. Terms that play a dual role (first and second function) have been discussed by Quine (1960), Castañeda (1966, 1967), Kiteley (1968) and Loar (1972). A fuller explanation of what is going on when a term functions in the second way will be offered in section III.

The third possible function is a pragmatic one. It is not special to *oratio obliqua*, and could be borne by any term, whatever its semantic role. A may intend H to read some *extra significance* into the fact that he selected the words 'coloured man'. To get this across he may adopt a special tone of voice, or accompany his utterance with a gesture. But he need not. His audience may be sensitive enough to get the intended message without that (e.g. they may know that A usually avoids the term, hence infer that he is satirizing people who use it). There is, then, a second kind of double duty that the utterance of the definite description may do. It may be simultaneously used and *paraded* as though it were being quoted. Let us call the third the 'parading' function. An unquoted term in *oratio obliqua* cannot *just* be paraded, it must be always performing either the first or second function (or both) as well. If A parades it when using it opaquely, he hints that *Sally* is the sort of person who uses it. He may also be implying that the words are passing through Sally's mind at that moment.

Nevertheless, even if it were true that Sally used those symbols to 'think with', in some sense, this fact does not need to be mentioned by one whose aim is to specify the *content* of her thought. Another subject, or perhaps Sally herself, could have thought the same thing without clothing it in English words. A similar point could be made about *saying* the same thing: if John says 'The book is green' and Pedro says 'El libro es verde', their words are different but the content is the same. In the case of linguistic symbols it is obvious that meaning or representational content is a different property from acoustic or orthographic *shape*. The distinction must apply equally to internal symbolic representation. Each token must be neurally instantiated in some form or other, but tokens of the same content type can be distinct 'symbols', if internal 'symbols' are individuated by their neurophysiological shape, or by any other criterion that is not functional-semantic (cf. Dennett, this vol.).

This point is crucial to what follows, so it deserves a little more elaboration. (Also the waters were muddied somewhat by Fodor's (1980) article, as can be seen from some of the commentaries and Fodor's replies.) Suppose we ask, is there really a sharp dividing line between *what was thought* and *how it was thought*? The question cannot be given a simple 'Yes–No' answer, since there are many kinds of 'how'. It depends what we mean by a 'way of thinking' a thought. To think of someone as a coloured man is a way of thinking, a 'mode of presentation' in Frege's terminology, and this is indeed a part of what is thought. But to ask *how* someone thought could be to ask for the mechanism, or the symbolic medium through which the act was performed, and here there is indeed a clear dividing line. One may know S's mode of presentation yet still be in the dark about the vehicle through which S encoded that mode of presentation. Conversely, knowing what the vehicle of representation looks like is not the same as knowing its internal, 'functional-semantic' role.[2]

Still, might there not be some cases, perhaps cases where S thinks 'in words', as we say, where the medium is an aspect of the content? I believe not. But two kinds of case occur to me as possible sources of confusion. One is where, for example, Sally thinks of someone as *describable as* a 'coloured man'. This complex relational mode of presentation is available only to a subject who has knowledge of a symbolic representation in English. But it still needs to be encoded inside Sally, and this will require that she have an internal representation of the English words 'coloured man'. Her internal symbol is distinct from that which it symbolizes, in this case as in every other case.

Another cause of confusion is the phenomenon of non-intersubstitutable synonyms. For example, it seems possible that Sally might think at t that Tom is an unmarried man yet not be *ipso facto* thinking that Tom is a bachelor. This is puzzling. Although being a bachelor just is being an unmarried man, there seem to be two distinct possible modes of presenting this property in thought. Some explanation is needed. I offer mine in section III below. However, we can see straightaway that one way of tackling the problem cannot be right. This is to hypothesize that Sally understands the words 'unmarried man' but does not understand 'bachelor', and that, as a consequence, she has an *unmarried man*-concept but lacks a *bachelor*-concept. The explanation is no good, for several reasons. It may not be true that Sally understands the English words 'unmarried man' (she might be a foreigner); or, conversely, it may not be true that she lacks understanding of 'bachelor'. More importantly, it may be that she possesses both an *unmarried man*-concept and a *bachelor*-concept (whether or not she knows English), yet she does not *exercise* her *bachelor*-concept when thinking her thought at t. The explanation erroneously equates exercising a concept with thinking of something as falling under a linguistic predicate.

To summarize the claims made so far: when A uses a definite description (or any other term) in the first or third ways, it is not helping to specify the content of S's thought. To do so it must function in the second way. Ascriptions which refer to the object of thought in one part of the clause, and which specify *part* of the content in another part of the clause, are bases for practically useful context-charged classifications, but the types are hybrids (cf. Fodor's notion of 'transparent taxonomy' (1980)). Fortunately for the would-be classifier or text-book writer, the content-specifying job is done by the sentence-type (relative to an interpretation in a language). His content-ascriptions can be read and understood by anyone who understands the language of ascription. Who the author was, and when and where he made the ascription, drop out as irrelevant, once it is clear what his ascribing-sentence means.

II. The limitations of specifications in *oratio obliqua*

In this section I aim to show that there are two large classes of thoughts whose contents cannot be accurately specified in *oratio obliqua* in a given language of ascription, e.g. present-day English.

(i) Indexical thoughts

Let us consider ascriptions whose 'that'-clauses contain personal pronouns, possessive adjectives, demonstrative pronouns, or indexical adverbs and pronouns like 'here', 'now', 'there', 'then', 'today', 'yesterday', etc. It is convenient to start by examining the roles of these words in the context of 'S says that . . .' and then to extend the findings to 'S thinks that . . .' It is clear that the pronouns have the three possible functions sketched above.

First, A may use a pronoun in *oratio obliqua* either to make a demonstrative reference or to hark back to an antecedently introduced term. When the pronoun is anaphoric, its antecedent may be the subject of the sentence in which it occurs, or it may be some other term. Syntactic ambiguity arises in sentences where there is more than one possible antecedent (e.g. 'Sally said to Mary that she was the prizewinner'). When it functions solely in these ways, the pronoun is not specifying any part of the content of S's utterance.

Secondly, provided the pronoun is anaphoric, it may also be construed opaquely, i.e. as not replaceable *salva veritate* by its antecedent or by any other term that denotes the same thing as the antecedent. It is, in Castañeda's (1967) terminology, a *quasi-indicator*. (As in 'The Editor of *Soul* believes that he* (i.e. he himself) is a millionaire'—Castañeda's (1966) example.) 'Quasi-indicators are the expressions which in *oratio obliqua* represent uses, perhaps only implicit, of indicators, i.e. uses which are ascribed to some person by means of a cognitive or linguistic verb.' (Castañeda 1967.) Castañeda *signals* them with an asterisk; one might *define* them as pronouns (or other indexical terms) in *oratio obliqua* that play a dual role (anaphoric and opaque).[3] Pronouns functioning in this way *do* (purport to) give information about the content of S's utterance.

Thirdly, A's utterance of a certain pronoun may serve to exhibit the very word that S used when she spoke, as if A were reporting in *oratio recta* (e.g. 'Sally, talking about her new stereo, said that *she* was a beauty'). It is important to realize that if the pronoun *qua* paraded conveys the information to H that Sally referred to something using the English feminine third person pronoun, it conveys that information *pragmatically*. It is easy to overlook this third function or to confuse it with the second, since a paraded pronoun is invariably being semantically used as well, perhaps as a quasi-indicator, and the grammar often dictates that A must use one particular pronoun. So it is hard for A to eke out any extra nuance from his choosing that pronoun rather than some other. However, there is a clear conceptual distinction between the content-specifying function and the parading function, as emphasized above.

Now consider the sentence 'Sally said that she was the prize-winner', where 'she' is anaphoric. If it is *just* anaphoric, and the only antecedent is 'Sally', we can infer that Sally referred to

Sally when she spoke. But the sentence refrains from giving any
clue as to what term she used. All we may conclude is that there
is some term 'α' denoting Sally (or believed by Sally to denote
Sally), such that Sally said 'α is the prize-winner' (assuming
Sally spoke in English). Let us assume, however, that 'she' is
opaque as well as anaphoric, i.e. it is a quasi-indicator. We still
cannot deduce how Sally spoke of herself. The quasi-indicator
implies that Sally referred to herself not by name or by
description, but by a pronoun. But it does not specify which
pronoun she used. Normally she would refer to herself in the first
person. But on this occasion she might have referred to herself in
the third person. In either case it remains true that 'she' is not
replaceable by any other term. Evidently, knowing that 'she' is
a quasi-indicator, as defined above, does not settle the question of
what personal perspective she adopted at the time of speaking.
Only one thing in practice settles this: resort to her actual words.

A similar underdetermination may exist when 'she' harks back
to an antecedent other than the subject of the main verb. In the
sentence 'Sam said to Mary that she was the prizewinner', 'she'
is the *oratio obliqua* substitute either for Sam's use of 'You'
(addressed to Mary), or to Sam's use of 'She' (referring to Mary).
It may seem hard to envisage a conversation in which Sam,
while addressing Mary, refers to Mary using 'She'. But it is not
all that hard. The situations in which this might occur are
parallel to those in which Sally might refer to herself using 'She'
(Castañeda 1966; Perry 1979). Generally, if S speaks of someone
in the third person when a first- or second-person stance is
appropriate, the reason is that S lacks a piece of contextual
information, hence does not *realize* that a first- or second-person
stance is appropriate.[4]

Since the opaque construal of 'she' does not discriminate S's
personal prospective, A's *oratio obliqua* does not convey what was
said from S's point of view. All it does is tell H that S referred to
an object using *some indexical expression or other*. Similar remarks
apply with respect to the temporal or spatial perspective adopted
by S when S refers to a time or a place indexically. They also
apply when S refers to an object by means of a description which
contains an indexical expression. For example, Sally is looking

at photographs of people's heads, including one of her own head taken from behind. She says, pointing to the girl in the picture, 'Her hair looks nice'. Had she realized that the girl was herself, she would have said 'My hair looks nice'. There is no way of discriminating between these in an *oratio obliqua* report of what she said which contains a quasi-indicator. The conscientious classifier of things said is hamstrung by the rule of English grammar which lays down that if a reporter uses a pronoun anaphorically, it must agree in person, number and gender with the antecedent, on pain of not achieving the desired back-reference.

I do not wish to undervalue the perspectival subtleties that *can* be conveyed in English *oratio obliqua*. For instance, practically all the temporal information that is conveyed by S's use of *tenses* can be captured in *oratio obliqua*, either by following the sequence of tense rules, or by specifying the time of S's utterance and reporting it with the help of a tenseless main verb: 'S says at t that . . .' The sort of temporal information that *is* lost in *oratio obliqua* is information about S's temporal perspective, where S refers to a time indexically using 'now' or 'then', 'today', 'tomorrow', etc. Similarly, if S makes a demonstrative reference to a place, A's report cannot convey the indexical spatial stance which S took up *vis-à-vis* that place.

It now remains for us to apply these findings to 'S thinks that . . .' The grammar is the same; the philosophical implications would be the same too, if thoughts were capable of being indexical in the ways that utterances can be indexical. The conclusion would be that thought-contents have perspectival aspects which are not picked up in *oratio obliqua*. I think this conclusion is basically correct. But the assumption that there is a full range of mental indexicals really requires defence. There are two high hurdles to cross.

It might be objected, first, that the notion of an indexical thought has no clear meaning. Most thoughts are never expressed, so there is no *oratio recta* to appeal to as criterion. If a thought is expressed and its expression contains an indexical, that fact does not necessarily prove that the thought itself is indexical. What might be happening is that S thinks non-indexically but chooses to express himself indexically in order to facilitate quick communication. The claim that some mental states are indexical is sheer metaphor.

This sweeping objection has sting; but it can be neutralized. In strict truth *all* talk about mental content is ultimately metaphorical. The question is, what aspects can we make literal sense of? I believe we should view the claim that there are indexical contents as a *theoretical* hypothesis, invoked within the framework of a theory shot through with analogy. This is, in effect, what Perry (1979) does. He treats the behavioural effects of a certain belief as relevant evidence of its being indexical. The *functional* difference between my 'self-locating' belief that I am making a mess and my descriptive belief that the F is making a mess (when I do not realise that the F is me) is naturally mirrored in the difference between the semantic function of 'I' and the function of the definite description. The analogy of role justifies and explicates the claim that the former belief is indexical. A further sort of evidence, not necessarily to be despised, is a sincere avowal by the subject that, as far as he can tell by introspection, he was not thinking of himself descriptively or by name or as a third person. It seems reasonable to conclude that the thought occurred to him in a distinctive, self-presentational mode.

But there is a second, more telling objection. Concede that there is a mental analogue of 'I'. The argument against *oratio obliqua* has been that it fails to capture *distinctions* between S's first-person indexical and S's second- and third-person indexicals. To carry the argument over to thoughts, one must show that there is in thought something corresponding to 'You' and 'He' and 'She' (and 'That' and 'This'), *and that these cannot be reduced* to descriptive modes of presentation plus the first-person mode of presentation. If the second- and third-person mental analogues were *analysable*, this would mean that all the necessary content distinctions could be captured in *oratio obliqua* after all. For consider an ascription such as 'S thinks that he is making a mess', where 'he' is opaque. If the only irreducible indexical thought-perspective is first-personal, then 'he' must represent S's mental analogue of 'I'. There is no alternative possibility. For suppose for the sake of argument that A's 'he' represented S's mental analogue of 'He'. This, according to hypothesis, is reducible to a descriptive mode of presentation (which may or may not include a mental 'me'). There must have been some term which specifies this mode of presentation, and which could replace the term 'He' *salva veritate*. But this contradicts the initial assumption that 'he' was opaque. Therefore, whenever a pronoun in *oratio obliqua* really is non-replaceable, it must be the mark of a first-person indexical thought, according to this view.

Furthermore (the objection continues), how could it possibly be proved that there is an irreducible mental analogue of the demonstrative uses of the pronouns 'You', 'He/She', 'This' and 'That'? The notion that S can demonstrate something outside himself purely by *thinking* comes perilously close to self-contradiction. Nor does the Perry-style argument offer any support here. In order for my belief about *you*, for example, to motivate me to move in your direction, I must mentally locate you in a me-relative way, e.g. as the person immediately to my left.

At first sight, the objection seems pretty strong. (The general line has been powerfully argued by Schiffer (1978).) To do it justice one needs to develop a full theory of indexical thought, a project beyond the scope of this paper. Nevertheless I think it is possible to extract a few points in support of my present thesis.

First, the idea of one mode of presentation being analysable in terms of another is by no means clear, particularly when conscious thoughts are at stake. Intuitively, it seems that if there really are two distinct modes of presentation, there must be two distinct contents. *Individuation* of contents is what we are interested in. Talk of reducibility relations between contents presupposes that they have been distinguished.

We must be clear about the different senses in which contents are *structured*. A concept may be present in S's mind at t either in a conscious, activated way or in a submerged way. For example, the thought strikes me at t that my cigar is alight. I think of the object in front of me as a cigar. I also thereby think of it as made of tobacco, given that my concept of a cigar is, in part, a concept of something made of tobacco. But that component of my concept was not separately exercised at t; if it had been, the thought would have been subjectively different. We may thus distinguish in a rough, intuitive way between the concept under which S consciously thinks of an object (the mode of presentation) and the subconscious conceptual ingredients of that mode of presentation. The latter are in one sense parts of the content of the thought, since they are ingredients of something which is part of the content. But they are not parts of the content in the same sense in which the mode of presentation is part of the content, since they do not present themselves as parts to S. If the relation between an indexical mode of presentation and its putative descriptive *analysans* is supposed to be anything like the relation between a conscious mode of presentation and its subconscious conceptual ingredients, then the claim that indexical thoughts are analysable is irrelevant to the present issue. For even if such analyses were available, they would not reduce indexical modes of presentation to descriptive *modes of presentation*.

So the objection really needs to be that there *are* no indexical modes of presentation (apart from the first-person mode). But this is surely false. We should not get hung up on the word 'demonstrative'; the crucial feature of indexical thought is that S lets the sensory context help pick out the object. For example, at t_1 S sees a man in a mirror and thinks at t_1 that he has a hole in his jacket. It may be that S *could* think of the man under various descriptive aspects, including self-related aspects (e.g. as the man whose jacket he is looking at, the only man in the mirror, and so on).[5] These are *potential* modes of presentation available to S. But S can surely think of him without *actualizing* any of these. S does not need cognition to pick a man out, because his current perceptions already pick one out.[6]

Of course, perceiving involves interpretation of the raw input. It is a process whose end-product has structure; a structure imposed by submerged conceptual ingredients. Nevertheless, perceiving an object and conceiving an object are functionally different modes of presentation, and the difference closely parallels that between demonstrating and describing. It would seem, then, that this is enough to make the present point. Let it be the case that the man in the mirror is S himself. An opaque *oratio obliqua* report of what he thought does not specify whether his mode of presentation was first-person or third-person ('demonstrative').

There is also such a thing as thinking of a *mental object*—a sensation, a mood, a thought, etc. The object or event may be either a type or a token. Suppose the object of thought is a particular pain, tied down to a particular time (or stretch of time). If the thought is concurrent with the pain-token, S's cognitive contact with it is peculiarly direct, though it is clear that concurrence alone is not sufficient to make the thought a thought about that token, and that some sort of causal connection is required. Alternatively, if the thought occurs after the pain, it may present the pain indexically via a memory-trace left by that token. Such thoughts might naturally be expressed by S's saying (e.g.) 'This is horrible' (at the time of the pain), or 'That was horrible' (afterwards). These have different *predicative* contents, so the difference between them can be captured in *oratio obliqua* provided the 'that'-clauses mirror the tense differences.

If S thinks that a certain pain-*type* is (tenselessly) horrible, he must present it in a conceptual mode, but he may descriptively identify the type via an indexical reference to a token of that type. For instance, he thinks of the same pain-type on two separate occasions, exploiting the token that occurred at t_1. At t_1 he mentally 'describes' it as 'the type of pain of which *this* is an instance', at t_2 as 'the type of pain of which *that* was an instance'. Again, *oratio obliqua* conversion *can* register the different temporal perspectives. Although both conversions would contain the same quasi-indicator 'it', the 'that'-clauses contain subordinate clauses whose verbs are, respectively, 'is' and 'was'.

However, even such intimate objects as these can be thought about from a variety of other indexical perspectives, in addition to the two temporal perspectives. People *address* their own mental states occasionally! Winston Churchill suffered from bouts of depression which he called his 'black dog'. (The same depression kept on coming back.) One can imagine Churchill mentally 'saying' to his black dog 'You have gone on too long'. The phenomenon is familiar enough; cf. Dr Jekyll and Mr Hyde, Christians fighting temptation ('Get thee behind me, Satan!'), case-histories of schizophrenia. It is beside the point to protest

that such personifying is merely a *façon de parler*. Of course it is.
But the internal perspectival differences exist, and the analogy
of internal dialogue provides the requisite distinctions. *Oratio
obliqua* is forced by grammar to obliterate them.

(ii) Thought-contents foreign to A

So far we have considered indexical thoughts about individuals.
There are also thoughts with completely general contents which
cannot be accurately specified in English *oratio obliqua*. The
nature of the difficulty is most clearly brought out by cases in
which S's native-language expression of the thought has no exact
English translation. The difficulty is not confined to English, of
course. For every thought whatever, there are possible ascribers
who will have trouble specifying its content in *oratio obliqua* in
their language (call it L_{AH}—the language of A and H).

For convenience of exposition, I stipulate that in the examples
in this section, the sentence which S would utter if he were to try
to express the thought in his native language (L_S) does in fact
express the content reasonably accurately, given its standard
sense in L_S. Later on, the connection between utterance and
thought will be scrutinized. Still, the condition is not wholly
unrealistic; many real-life examples satisfy it.

Consider, then, a Spanish-speaking S who expresses what he is
thinking by saying 'Todos los ingleses son rubios'. In Spanish–
English dictionaries, 'rubio' is commonly translated as *fair-haired*.
In fact, Spaniards apply the term more widely to cover the range
from blond to fairly dark brown. How should A specify this
thought-content in English? It is not strictly true to say that S
thinks that all English people are fair-haired. If you compare S
with a sample of subjects who really do think this, you will find
that S reacts differently to certain bits of counter-evidence. For
instance, he is not disposed to retract his thought upon meeting
a succession of brunettes whom he thinks are English. Controlled
tests would reveal that his cognitive state played a role in
inference and in the causation of behaviour which was not the
role to be expected of a thought that all English people are fair-
haired.

On the other hand, it is not right to say that S is thinking at t that all English people have hair whose colour is within the range from blond to fairly dark brown. This ascription does not accurately specify how he conceptualized the hair-colour in question. Given that S belongs to a culture where they treat it as a basic hair-colour category, S's concept is not so readily *analysable*. His concept is structurally and functionally unlike a *colour-within-the-range-blond-to-quite-dark-brown* concept. Thus the English ascriber can have good reason to conclude that his 'that'-clause does not capture the content as accurately as a Spanish *oratio obliqua* ascription would. Hence he cannot maintain that his own ascription is an accurate *specification*. *No* ascription in English can meet the high standards of precision required.

Inaccuracy can arise if S speaks a different dialect of English from A. For instance, there is (I believe) no equivalent in Chaucer's English for the word 'sophisticated' in the sentence 'Town-dwellers are more sophisticated than peasants', uttered by me. Chaucer could not have specified exactly what I think when I think the content expressed by that sentence. No medieval paraphrase could catch the twentieth-century nuances. (If I am mistaken about this example, there are plenty of others.) Just as A may be unable to *say* what the content of S's thought is, so may he lack the concepts to *understand* S's thought-content. He can remedy this ignorance provided that S is not too alien. The obvious steps are to learn S's language, live in S's culture, and learn to think like S. In cases like these, where my stipulated condition holds, the task of specifying S's thought-content just *is* the task of specifying the intension of the expressing-sentence. To discharge the task using *oratio obliqua*, A must find a sentence of L_{AH} that has the same intension (and the same intensional substructure) as the sentence of L_S. Since the intension of a sentence (or term) in a language depends upon the totality of its logical relations to other sentences (terms) in that language, perfect translation demands that L_S and L_{AH} be intensionally isomorphic in the relevant areas. Probably no two languages are perfectly intensionally isomorphic. There are sure to be sentences of one that cannot be exactly translated into the other; indeed, no sentence can be exactly translated if we assume a thoroughly holistic view of intension. The greater the overall divergence between languages, the looser the translation.[7]

The conclusion is, therefore, that a specifier of contents who sticks to *oratio obliqua* has no hope of achieving his goal of *exactness* if the subject is not from the same linguistic and cultural community as himself. He must reconcile himself to this crippling limitation, despite his believing that foreigners' thoughts indeed *have* specifiable contents, and that foreign psychologists could probably specify them.

But surely there is something absurd about this situation. A psychologist should not have to renounce all attempts to specify what S thinks merely on the grounds that S thinks differently from himself. He ought to try to adopt the naturalistic stance of a disinterested observer, not the stance of a socializer. So if being a compatriot is a condition of the possibility of *being accurate* in *oratio obliqua*, the serious taxonomist has no choice but to move out of *oratio obliqua*.

III. Specifying contents by modelling them on sentence-intensions.

This section outlines a method of specification which uses *quoted* sentences. This method, already hinted at, is perhaps the most natural way of handling the above difficulties without straying away from the intuitive notion of content. However, specifying content in the new style is a different kind of enterprise from content-ascribing, although there is a bridge that links them. The nature of the difference will be explored in this section. It is a difference between two stances that a classifier can adopt *vis-à-vis* S: the stance of a fellow-subject, which one might call 'first-order psychological', and the stance of 'second-order psychology', from which the subject is viewed more from a distance, as an entity standing in a relation to a language-system.

The basic idea behind the new method is expounded in Sellars (1967); Aquila (1977); Churchland (1979); and hinted at in Leeds (1979).[8] The latter two note the striking similarity between content-fixing and *measurement*. Ascribing content is like plotting inner states on a graph, except that the scales are not numerical and the states are not arranged along *dimensions* in any obvious sense. Minds, being holistic, demand a different kind of mapping-technique.

The key assumptions are that content-endowing discourse is *analogical* (Aquinas; Sellars 1956, 1967; Geach 1957), and that the analogy is illuminated by a *functionalist* view of the mind (Putnam 1960, 1966; Fodor 1965, 1968; Block and Fodor 1972; Harman 1973; Block 1978). Functionalists hold that mental states are internal states of an information-processing system, type-individuated in terms of their functional relations to input, output and one another. They are divisible into functionally defined *genera* (e.g. motivational, cognitive, affective), these being further subdivisible into species and subspecies. The division of states into *content*-types marks a more abstract functional classification. Each state in the 'cognitive' category, for example, has a characteristic profile of potential interactions with other cognitive states when S is engaged in *reasoning*. We regard the role that it has in inferences as the role appropriate to, and (partially) definitive of, a state having a certain 'content'. For instance, one tiny fragment of role-information about the type we call 'thinking that all cows are mammals' is that if it is instantiated in S at the same time that the state we call 'thinking that all mammals have teeth' is instantiated, and if S is engaged in deductive reasoning at that time, then, provided certain other conditions hold, S will enter the state we call 'thinking that all cows have teeth'. The 'other conditions' pertain to motivation, attention, maintenance of normal functioning, absence of external interference, etc. Whether S ever actually instantiates the state in question, or goes through this sequence, will depend upon a host of other factors that are not our present concern. The important thing is that the state-type is defined by potential transitions of this sort (and by potential transitions in practical reasoning too).

The hypothesis now under consideration is that a state is deemed to have content only when a certain condition holds. Not all S's psychological (behaviour-controlling) states are contentful states. The condition is that the psychological state in question occupies a position in a network of states whose characteristic interrelations in S *qua reasoner* are matched by the logico-semantic relations between the sentences of a language. It is a contingent matter whether there exists a stratum of psychological states in S on which such isomorphism is to be found. It depends upon what S is like, and also upon what languages are available. But when a match can be set up, it can be exploited. Each of S's states on that psychological level can be assigned a sentence as its *label* or *identification-mark* within that system of mapping. There is no need to *describe* its role in reasoning in tedious detail. Using the language as a model, you can identify a state simply by citing its *label*.

Anyone who understands the modelling-language (and who thereby knows the intension of the label) can work out how the state interacts in reasoning processes with other states that are assigned labelling-sentences. So the knowledge that such and such a label has been assigned has explanatory and predictive value for one who is in a position to exploit the model. It is important to remember that the model is the intensional structure of a whole language. What it models is the whole of S's cognitive structure. The content of a particular thought is identified, within the framework provided by a global modelling *of* S *on* L (as we may put it), as the *intension* of a particular sentence in L. The sentence itself serves as an index, recognizable to A and H by its sound or shape, which happens to mark that intension (cf. the device of 'dot-quoting' invented by Sellars (1967, pp. 311ff)). Contents may be further subdivided into 'content-elements' whose roles in S's cognitive system are analogous to the intensional roles of words and phrases in a language-system. When S is modelled on L, words and phrases in L can serve as labels of those content-elements. The present hypothesis is that the mental items we call 'concepts' and 'modes of presentation' are content-elements discerned through the perspective of such a model.

On any view, the intension of a sentence is intimately related to the intensions of its parts. But there is an exceptionally *pointilliste* way of individuating sentence-intensions whereby the intensional substructure of a sentence is not merely determinative but *constitutive*. On such a taxonomy, two sentences can differ in intension even if they have the same truth-condition. (Field calls it 'fine-grained intension' (1977).) To be co-intensional, they must be built up in the same way out of co-intensional parts. Even such fine distinctions as these can be exploited, provided there exist corresponding real differentiations in the psychological states of S.

The idea of modelling thought-contents on intensions provides a *rational reconstruction* of what it is to ascribe content in *oratio obliqua*. When A is engaged in first-order psychological interpretation, A is actually exploiting *his own language as a modelling-language for S*. Within the context of this model, the sentence in A's 'that'-clause functions as the public index of S's thought-content. In the act of uttering the ascribing-sentence, A *exhibits* that index to H, but A does not say that it is an index nor intend it *as* an index necessarily, nor does he say that he is viewing S through a model. A *uses* the embedded sentence as an instrument of communication, expecting and intending H to grasp its sense, in just the same way that he would if he uttered it unembedded as the expression of a thought of his own. To understand it, H has to decode it. The real-time decoding procedure transforms H's cognitive state so that he comes at a certain stage in the process to apprehend the same content for himself. H does not stop there, of course; the end comes when H has decoded the whole of A's utterance fully, and has entertained the thought that S thinks that p, rather than the thought that p. Understanding is, one might say, a way of knowing content 'by acquaintance'. Borrowing from Davidson's idea of 'samesaying' (1968), we could describe it thus: A, by encoding, and H by decoding, fleetingly become *samethinkers* with S, if what A says about S is true. This does not mean they become *samebelievers* (see Stich, this volume). A and H need not subscribe to S's *attitude* in order to apprehend S's thought-content.

A pre-condition of reasonably accurate communication of S's content is that A and H should have cognitive structures similar to S's, and that the contents of all three should map reasonably well on to the intensions of sentences in L_{AH}. The precondition is likely to be satisfied if $L_S = L_{AH}$. When all three think alike, A and H operate from within the model, unaware (usually) that they are viewing S through the perspective of a *model* at all.

Naturally, they must know the rules of the ascribing-game. But they need no explicit grasp of the fact that the game requires them to think and speak analogically. For this reason the suggested reconstruction is not proposed as an analysis of the *meaning* of sentences of the form 'S thinks that p'. Any formula which makes the analogizing explicit will surely mean something different.

The sketch I have just given is extremely brief. Many details remain obscure and unexplained. But there is enough there, I hope, to give the flavour of a rationale for the new style of content-specification. For if it really is the case that content-ascribing involves modelling, then a philosophically sophisticated A can become aware of this fact. He can see himself as having a choice of two content-specifying stances to adopt *vis-à-vis* S. He can characterize S's mental contents either *through* the model provided by L_{AH}, or *as related to* the model. The latter requires that he distance himself from L_{AH} in its capacity as model. He now conceives of sentences of his own language as *objects*, and he surveys the way they map on to S. To describe the scene, he must *refer* to those objects. The hallmarks of the second loftier stance are, therefore, that A *cites* a sentence of the modelling-language, and he says or implies that the language is a modelling-language. So he ceases to say (e.g.) 'S thinks that snow is white', and says instead 'Within the model of S that my language provides, the content of S's thought is the intension of "Snow is white"'. It is only one small further step to replace the token-reflexive phrase 'my language' by the name of the language. The specification then meets the standards of impersonality appropriate to scientific discourse. This form of specification, cumbersome though it may be, is more perspicuous than the *quasi oratio recta* locution favoured by Geach (1957): '*S* thinks "Snow is white"'. This locution is itself metaphorical: it avails itself of an analogy between thinking and speaking without revealing the basis of that analogy. The great virtue of a second-order specification is that it makes it *explicit* that the thought-content is being identified in terms of an analogue.

I can now explain better what I mean by 'the stance of second-order psychology'. First-order and second-order psychology differ neither in subject-matter nor in goal. They both aim to individuate and catalogue psychological states whose identity-conditions are fixed by their contents-for-S. Whichever stance the psychologist adopts, his conception of the task derives from his general mastery of discourse about other minds, together with the concomitant belief that there is something that it is like to be S. He will have before him certain paradigms of psychological insight. Novels, films, and conversations with friends will have afforded opportunities to identify with other characters. He knows that there exist ways of thinking of which he has no inkling, but he knows also that there exist perspectives other than his own that he can think himself into quite readily.

The difference between the two stances lies only in this: that the first-order psychologist tries to specify content like a novelist, by re-expressing it; the second-order psychologist is content to describe it without re-expressing it. If successful, both can be said to know what the content is. The former knows it by acquaintance, the latter by description.

Adoption of the second-order stance obviously widens the range of contents to which A has specificatory access. For A may be in a position to know that there is a language L and a sentence Z in L, such that the intension of Z is the analogue of a certain thought-content in S, even though A does not understand a single sentence of L. This piece of information can be communicated to hearers who do not understand L. Thus there is a sense in which A and H can know what the thought-content is like, despite their ignorance of what it is like to think it.

As we saw in the 'rubio' case, *oratio obliqua* ascriptions are inaccurate when S thinks differently from A and H because the language of ascription is unable to serve as an accurate modelling-language for S. The obvious remedy in that case is to switch to the second-order stance and specify S's content under the description 'the intension of "Todos los ingleses son rubios", relative to a modelling of S on to Spanish'. Inevitably the specifier sacrifices something, for he can no longer officially aim to invoke in H the experience of same-thinking with S. But he must decide what line of business he is in. Does he want a more accurate statement of what S thought, or does he want an inaccurate Anglomorphic rendition? It is worth pointing out, however, that the descriptive specification does not sever the connection with first-order psychology altogether. Since it mentions Z and L, it hints at the procedure one should follow if one wished to report the thought accurately in *oratio obliqua*. In this case the recipe is: learn Spanish.

We may lay it down as a general rule that in a case where S expresses his thought with precision in his native language, A can specify its content precisely by referring to the intension of S's expressing-sentence. In cases where the 'perfect expression' condition does not hold, it could still be A's best bet to model S on L_S, because L_S might be the best model available. (See below for further remarks about imperfect modelling). However, it need not necessarily be the case that L_S is the best model. There could in principle be a language unknown to S, the sentences of which map perfectly on to S's thoughts. A second-order specification that models S on language L certainly does not entail that S understands L. Any theory which insisted upon such an entailment would be wrong, for the possibility must be left open that S does not speak any language at all.[9]

If there existed two intensionally isomorphic modelling-languages for S, A would be free to use sentences of either language as labels (cf. Sellarsian 'dot-quoting' again (Sellars 1967, p. 311)). Suppose for the sake of argument that German and English are two such languages, and that 'The moon is round' is a perfect (intension-preserving) translation of 'Der Mond ist Rund'. One and the same thought-content could then be specified under two distinct descriptions. Because our second-order specifications explicitly equate contents with intensions, of which the cited sentences are merely extrinsic labels, they are immune to the Translation Problem posed by Church (1950).[10]

How are we to deal with the problem of apparently synonymous predicates that cannot be intersubstituted in an *oratio obliqua* ascription? Suppose that English has two predicates, represented by 'P_1' and 'P_2', that are fine-grained intensionally equivalent. Then, necessarily, any thought whose content is the intension of 'All P_1's are neurotic' is a thought whose content is the intension of 'All P_2's are neurotic'. The orthographic difference marks no content difference. Yet it seemed possible for Sally to think at t that all unmarried men are neurotic, without thereby thinking that all bachelors are neurotic. There are only two possible responses: either we concede the possibility and argue that 'bachelor' and 'unmarried man' are not fine-grained cointensional, (i.e. not synonymous for modelling purposes), or we deny the possibility. If the latter, we incur an obligation to explain how it could have seemed possible. One line might be: it seems possible only to those who first imagine a background context of A's utterance in which it is for some reason *odd* or *infelicitous* for A to say 'Sally thinks that all bachelors are neurotic', and who then confuse infelicity for falsity. But this explanation is not very convincing. I prefer the first response: 'unmarried men' has an intensional substructure that 'bachelors' lacks, and Sally's mode of presentation at t actually possessed that substructure.

I want to stress again that it is *not* claimed that the English phrase 'unmarried men' (or a 'neural token' thereof) is an internal constituent of Sally's thought at t. It belongs to the model, not to that which is modelled. In a second-order specification, the quoted words function simply as labels tagged on by the specifier-*cum*-theorist. The notion of a modelling-language is quite distinct from Fodor's 'language of thought' notion (Fodor 1975, 1980). S's 'language of thought' is definable as 'the medium of internal representation in which S conducts his thinking'. Fodor encourages us to regard it as a system of cerebral symbols. S 'uses' them to think with, and 'translates' them into public symbols when he talks. But we cannot, on pain of vicious regress, say that S understands what they mean, or that he intentionally picks this symbol rather than that. Fodor may not find this confusing, but a lot of other people certainly do. Talk in these terms is really a misleading extended metaphor. Instead of dissecting the analogy, Fodor builds on it, encapsulates it, literalizes it.

Concerning indexical thoughts, the second-order specifier has the resources to distinguish as many indexical modes of presentation as are expressible in the modelling-language he chooses to employ. For example, if English is a suitable model of

Sheila, who thinks at t_1 that she is about to be bitten by the dog that has just come into view, her indexical perspective could probably be pinned down pretty accurately. There are many alternative possibilities here, each having its own distinctive explanatory potential. One possibility is that the content of her thought was the intension of 'That dog is about to bite her'— Sheila saw the scene through a mirror not realizing the woman was herself; and she thought of the dog demonstratively, while recognizing it to be a dog. This thought is one that Sheila could think on other occasions in the presence of other women and other dogs, seen either through mirrors or not through mirrors.

When S thinks indexically of presented objects, the nature of the perceptual experience will vary from case to case. I am inclined to regard these sensory differences as irrelevant to the strictly *cognitive* content. If we were aiming for a taxonomy of *experiences*, then qualitative differences between percepts would become crucial, as would differences of type among mental images, feelings, bodily sensations, moods and so on. However, this is not the sort of issue that can be settled *a priori*, by consulting definitions. It may be that many cognitive processes and states are 'non-discursive' and 'imagistic', in which case the taxonomy that is based on language-modelling will turn out to be quite inadequate. This is an area of current controversy among psychologists. Still, it must be acknowledged that a good case can be made, on purely *philosophical* grounds, for distinguishing between mental indexicals that reach out through different sensory modalities. Thus there might be one mental analogue of 'This' (or 'That') which picks out an object visually, another which picks out something aurally, and so on. (See Bach, this volume, concerning the argument of Schiffer (1978)). No public language, as far as I am aware, contains demonstratives associated with separate sense-organs. It would appear, then, that second-order specifications are not as fine-grained as the scrupulous taxonomist would wish, as far as indexical thoughts are concerned. The modelling-language needs to be supplemented with a vocabulary of sensory-modality markers. Then it becomes possible to represent S as thinking two thoughts simultaneously, the content of one being the intension of 'This (visual) is F', the content of the other being the intension of 'This (tactual) is not-F'.

Although indexical thought-contents are not completely conceptualized (cf. Burge (1977) on *de re* beliefs), they are nevertheless *complete* in the same sense that the intension of a sentence containing an indexical is complete. No thought-content can in itself be incomplete; anything less than a complete content is a content-*element*. However, a thought-content can be incompletely *specified*, by a formula that cites an unsaturated expression (e.g. a predicate on its own), or by a formula that quotes a sentence containing an *anaphoric* pronoun. Thus the theory of content cannot allow the following second-order specification to count as complete, where 'It' and 'her' are both anaphoric: 'The content of Sheila's thought at t_2 is the intension of "It is now six inches from her ankle"'. What is in Sheila's mind at t_2 is essentially a continuation of her thought at t_1. In context it is a complete thought. But its content is completed by being thought against the background of the earlier thought. In order to specify its content, one must specify the antecedent content-elements to which it harks back, and also specify *that it is* anaphoric. It differs in content from an indexical thought whose content is the intension of 'That is now six inches from her ankle' (where 'That' and 'her' are both demonstratives).

Clearly it is of taxonomic importance not to confuse anaphoric thoughts with indexical thoughts about previous thoughts. Both involve memory, but memory operates in many different ways. Mental anaphora requires that earlier content-elements be kept activated in the specious present. Short-term memory is involved; but S does not *remember* the earlier content-elements in any sense which implies that S thinks about them as objects, either conceptually or indexically. Sheila keeps her attention fixed on whatever the antecedent content-elements were about. Mental anaphora, like verbal anaphora, has no essential connection with demonstration at all. *A fortiori*, it has nothing to do with mental demonstration of events in the past; *a fortiori* it has no special connection with mental demonstration of mental events in the past.

In the remainder of this section I want to consider the question of *accuracy* in second-order specifications. How accurately can the specifier pinpoint the true content of a thought? To answer this we must be careful to distinguish between practically attainable accuracy and the kind of accuracy that is possible in principle. I believe that it is unrealistic to expect perfect accuracy in practice, but that from the second-order stance one can form a conception of what a perfectly accurate specification would be. This conception serves as a Kantian 'idea of reason' fixing the direction we should move in if we wish to make our actual specification more accurate.

The modelling of thought on language involves scaling the psychological structure of the individual against something public. The intensional structure of a language is determined by the rules of correct use that are created and recognized by a social community. Inevitably there will be individual psychological variations among people within a linguistic community. Consequently it is inevitable that the language will not be a perfect model of the cognitive structure of every native speaker. Some thought-contents are sure to slip through the social net. Indeed, it is probably safe to say that no individual's conceptual framework is exactly isomorphic with the intensional structure of his or her native language.

We may gain a sense of why this is so by considering what perfect native-language modelling would involve. The following thought-experiment conveys the flavour of it (though the picture is, for various reasons, oversimplified). Let S be an articulate, adult speaker of English who expresses his every thought. Suppose further that he is perfectly candid, perfectly self-aware, never self-deluding, and that he never expresses himself ironically or figuratively. The sentences he produces are, in effect, ready-made labels of his thoughts. But are they the *correct* labels? He utters the sentences intending them to be taken literally, but *which* sentence he selects to express a given thought depends on what he believes its standard intension to be. No normal human being has complete mastery of the intension of every word he uses. If there is any term in his vocabulary which he misunderstands or incompletely understands, the utterances of sentences containing that term will not accurately externalize the true contents of the relevant thoughts.

Perfect modelling requires, in addition, that all his *possible* thoughts should map on to the sentences of English. This necessitates (a) that every concept expressible in English be a concept that he has, (b) that he be incapable of having any thought whose content is not expressible in English. None of his actual thoughts will be correctly labelled unless this condition is met, because contents and intensions are determined holistically. If the intensional structure of L is more articulated than S's conceptual framework, content-specifications relativized to L will be too fine-grained to capture any of S's thought-contents accurately. If L is less articulated than S's conceptual framework, all specifications relativized to L will be too crude. However, it is quite unrealistic to expect that this condition will be met.

A better global match can be achieved if we take S's *idiolect* as the model, defining 'idiolect' in such a way that words of English that S does not know are not in his idiolect. Not even this will guarantee isomorphism, however, for S still has to understand perfectly every word in his idiolect. An idiolect is, after all, a *public* language. There is an ever-present possibility that S will express himself incorrectly in his idiolect.

No advantage is gained by defining 'idiolect' in such a way that the sentences of S's idiolect mean just what S means. The whole point of modelling is that the logico-semantic roles of labelling-sentences are determined by linguistic conventions that are independent of S. If they were not determinable independently, they would fail to provide the desired line of access.

We can be inductively certain that the cognitive structure of an individual will never, in fact, be a perfect mirror of the intensional structure of his idiolect. We can merely envisage this as an ideal possibility. Even if the possibility were realized, the perfect match would be unlikely to last long. The individual's repertoire of possible thoughts changes as he acquires new concepts and loses old ones. To keep step with the changes while preserving isomorphism, the model-idiolect would have to change its intensional structure continually, and we know that language-changes do not march in time with individual changes.

We may draw the general conclusion that every particular second-order specification that models S on his native language (or on any other public language), is likely to be inaccurate to some degree. It should be noted, though, that the fact that every subject occasionally makes mistakes of reasoning is *not* a proof that his thought cannot be accurately modelled on language. Such mistakes are mostly executive failures, errors in the way thought-types are instantiated in the course of a particular sequence. They are detectable and rectifiable by S himself.

Does the practical inevitability of inaccuracy mean that all available second-order specifications are probably, strictly speaking, false? I think not. We must distinguish two kinds of inadequacy from which they may suffer. The first is due solely to imperfect match between S and the given modelling-language L. For example, every attempt to specify what Pedro the ideal Spaniard thinks, relative to a modelling on English, will be inferior to some specification that models him on Spanish. A real-life classifier may not realize this, so he may propose a certain specification based on the model of English which is, in fact, the best specification he can get within the limits of that model. Naturally he will discard it if he finds a better model. Every specification based on a poor model is defeasible by a specification relativized to a better model. But it is not *false* to say 'Relative to a modelling of Pedro on to English, the content is the intension of sentence Z in English', if Z clearly the best label that English can provide. However there will often be a problem about determining which sentence *is* the best label. In the Pedro case, it is perhaps impossible to decide whether 'All English people are fair-haired' is better or worse than 'All English people have hair colour within the range from blond to fairly dark brown'. Substitute either of these for 'Z' and you get a specification whose truth-value is indeterminate.

The second kind of inadequacy arises when, in the context of an imperfect modelling on L, a label is assigned which is clearly inferior to some other sentence in L. In that case the specification is false. Pedro's thought-content at t, for instance, was not the intension of 'All English people have black hair' relative to the modelling on English, even allowing for the shortcomings of this model.

These remarks indicate that it is always risky to judge, of two people whose thought-contents have been only roughly specified, that they thought the *same* content. Interpersonal comparison cannot begin unless S_1 and S_2 are assessed on the same measuring-scale. So if S_1 was originally modelled on L_1, and S_2 on L_2, commensurability must first be established either by translating the labelling-sentence of one language into a sentence of the other (and thus perhaps compounding the inaccuracy), or by starting afresh using just one model, say L_1. It is unlikely that two distinct subjects will have exactly the same cognitive structure; if they do differ in this respect, the modelling on L_1 is sure to be imperfect for at least one of the subjects. More probably it will be imperfect for both, but more imperfect for one than it is for the other. Suppose, however, that both thoughts are eventually assigned the same labelling-sentence in L_1. Do they have the same content? Within the limits of accuracy of the model, they do. But the claim that they do will invariably be defeasible by a thought-experiment. If S_1 and S_2 really do have different cognitive structures, then no thought of one can really be content-identical with any thought belonging to the other. Situations that reveal the difference can be (hypothetically) set up, even if there is no common modelling-language superior to L_1 in which S_1's and S_2's thoughts can be assigned different labels. I believe that if we view inter-subject comparisons in this light, we gain insight into the fundamental difficulty—some might say defect—inherent in content-taxonomy. It is that the taxonomist has to trade off the benefits of precise specifications for S against the disadvantage of not being able to use the *same system of measurement* for anyone else.

The classification of objects by size, weight, age, etc. presents no such difficulty. If you wish to compare a British shoe and an American shoe for size, you can do so by measuring them directly in (e.g.) centimetres and millimetres. It is not necessary *first* to specify their sizes in relation to their respective national shoe-scales, and *then* compare the U.K. scale against the U.S. scale. Nor do you have to specify the size of one shoe in terms of the national scale appropriate to the other shoe, and then see if they are the same relative to that scale. You can forget about the U.K. specification and the U.S. specification, and also bypass any inaccuracies that those specifications may have involved.

I think we can also see how it might come about that, given two subjects who are cognitive replicas, and who think thoughts that really are content-identical, a classifier mistakenly concludes that the thoughts *differ* in content. This can happen if S_1 is imperfectly modelled on L_1, S_2 is imperfectly modelled on L_2, and the classifier knows that L_1 is not intensionally isomorphic with L_2. Suppose the labelling sentences are Z_1 in L_1 and Z_2 in L_2. Armed with two incommensurable specifications, the classifier follows the standard procedure for facilitating a comparison. He tries to translate Z_1 as best he can into L_2. But he knows that Z_2 is not a good translation. He concludes that, relative to a modelling on L_2, the content of S_1's thought is not the intension of Z_2, hence not the same as the content of S_2's thought. But S_1 and S_2 were, *ex hypothesi*, in the *same* cognitive state. What went wrong? The fault lay in the indirect procedure by which he brought S_1 and S_2 into comparability. The modelling on L_1 was imperfect, the translation of Z_1 was problematic, and the two inaccuracies added up. A better procedure would have been to forget about L_1 and Z_1, and to specify S_1's thought-content afresh, using the model of L_2 (or, better still, using another language that models both subjects better). This seems to me the right way to interpret the ingenious Twin-Earth experiments devised by Burge (1979a).[11]

In view of the practical certainty that every available second-order specification is inaccurate, one might doubt whether the scientific taxonomist gains a great deal by taking up the second-order stance. A small increase in precision is sometimes achievable in cases where S thinks very differently from the taxonomist. But as far as conceptual content is concerned, specifications that model S on L_S look unimpressive when one remembers how far from perfect that model is. Would it not be more sensible in the circumstances to stick with *oratio obliqua* and to make piecemeal qualifications here and there as the need arises? This is the way we normally express fine conceptual differences within first-order psychology. We say things like this: Simon and Sam both think that there is a cigar in the ashtray, but Simon's cigar-concept is a bit different from Sam's. Sam thinks of a cigar as something made of tobacco, but not as something carcinogenic; Simon thinks of a cigar as something made of tobacco and therefore carcinogenic. Their modes of presentation have different submerged conceptual ingredients. Also it would not be difficult to tag additional indexical perspective-indicators on to quasi-indicators in *oratio obliqua*. Second-order specifications are going to need such technical devices too.

I think we must grant all this. If the classifier's motive for switching to second-order specifications is a desire to achieve scientific precision, his efforts are misguided (cf. Field 1978). Nevertheless, it would be wrong to condemn the move to the higher stance as a pointless exercise. Its great virtue is that it exposes the true nature of the project. The first insight is that content-classifications are model-relative, the model is always imperfect, therefore the classifications are inherently imprecise. The second insight is that there exists a variety of incommensurable models, some better for some subjects, others better for other subjects, but no single model that is best for all subjects. 'Different model' entails 'different taxonomy', hence there is no single best content-taxonomy. Scientific psychologists who desire one are doomed to wait in vain. It is of some interest that this conclusion can be reached from a standpoint that is methodologically solipsist throughout, particularly in light of the fact that conclusions in somewhat similar vein have been held, by some philosophers, to depend upon a *non-solipsistic* conception of content. For example, it was argued by Stich (1978b) and by Fodor (1980) that contentful states are not suitable for incorporation into scientific psychology if the notion of content is tied to the notions of reference and truth. We can now see that content-based *classification* is unsuitable for strict science even if the contents in question are solipsistic. Systems whose internal states are modellable upon human languages are, of course, studiable scientifically, and are of special interest to psychologists, even if, from an A.I. point of view, they are but a small subclass of the total range of systems that could be called intelligent.

However, the fact that those states happen to be specifiable with the help of a model is of no abiding interest to psychology or A.I. viewed *sub specie aeternitatis*. Being real, the states and their intra-systemic functional roles are in principle describable directly, without the intermediary of a model. For strictly scientific purposes, literal descriptions are better describers of internal reality than analogical descriptions.

Notes to chapter 6

[1] 'The expression "apprehend" is as metaphorical as "content of consciousness". The nature of language does not permit anything else. What I hold in my hand can certainly be regarded as the content of my hand but is all the same the content of my hand in quite a different way from the bones and muscles of which it is made and their tensions, and is much more extraneous to it than they are.' (Frege 1956, p. 307.)

[2] The following passage from Spike Milligan's novel *Puckoon* (pp. 24–5) nicely satirizes the 'internal blackboard' metaphor:

'Together the two men sat in silence; sometimes they stood in silence which after all is sitting in silence only higher up. An occasional signal of smoke escaped from the bowl and scurried towards heaven. "Now Milligan," the priest eventually said, "what is the purpose of this visit?" Milligan knew that this was, as the Spaniards say, "El Momento de la Verdad", mind you, he didn't think it in Spanish, but if he had, that's what it would have looked like.'

(Thanks to Adam Morton for finding this.)

[3] This would amount to treating Castañeda's principle Q.2 as a definition: 'Quasi-indicators have necessarily an antecedent to which they refer back, but they are not replaceable by their antecedents.' (1967, p. 93.) Castañeda himself does not do this; he defines them implicitly in terms of the whole set of Q-principles.

[4] Castañeda (1967, p. 100) notes that there is ambiguity in such sentences as 'John told Paul that he* had been made full professor'. The present point is that even *after* it has been made clear which name is the intended antecedent of 'he*' (and whichever one is it), the quasi-indicator is still compatible with two distinct perspectival stances that John may have taken up. In fact, there will have been more than two possible stances. Imagine Queen Victoria muttering about herself. She might say 'I am hot', or 'We are hot', 'You are hot' (talking to herself), 'She is hot' (schizoid detachment), or 'One is hot' (the Royal 'one'). In every case she is saying that she* is hot.

Possibly Castañeda himself would prefer to define 'he*' and 'she*' more restrictively, as the *oratio obliqua* modes of conveying just *first-person* perspective. He writes 'The use of the pronoun "he*" cannot be analyzed in terms of the demonstrative reference of the strictly third-person pronoun "he". The only demonstrative reference of "he*" is bound up with that pertaining to the first-person pronoun "I".' (1966, p. 144.) In the later paper, however, he allows that 'he*' corresponds to demonstrative 'He' and 'You' as well as to 'I'. (See the chart under Q.5 in Castañeda 1967, p. 95.) This is merely a terminological issue. The important point is that S's personal perspective cannot automatically be retrieved from an opaque *oratio obliqua* report, because two or more *oratio recta* reports may have to receive the same *oratio obliqua* conversion.

[5] These modes of presentation are available to S only if he has the requisite concepts. That such a proviso has to be made constitutes an objection against the Schifferian

descriptivist position. Animals and infants can surely think about objects in the environment without having to be credited with the *concepts* of *self, looking at*, etc. (See also Bach, this volume.)

[6] There are fascinating problems about how to characterize the changes in S's judgements as he tries to keep track of objects over time. For example, if S closes his eyes after t_1, believes at t_2 that the man is still in front of him, and thinks at t_2 that the man currently in front of him has a hole in his jacket, his mode of presentation of the man is obviously different from before. It is a descriptive mode of presentation which contains indexical elements (mental analogues of 'now' and 'me'). Now suppose he thinks at t_2 that *that* man had a hole in his jacket, thinking indexically of the man via a memory-trace caused by his earlier perception. This too is a different mode of presentation from before (although I think both involve what Aquinas termed *conversio ad phantasmata*). The content of a perceptual thought cannot be retained after the percept has gone. Continuity is achieved by the psychological analogue of anaphora (see section 111 below). For more on the links with Aquinas, see Geach (1957).

[7] Scepticism about translation is a familiar theme in Quine (Quine 1960 being the *locus classicus*). But Quine holds that the very idea of determinate translation makes no sense; he eschews intensions and synonymy altogether; he claims that the content of a belief is intrinsically indeterminate. My position is that exact translation is possible in principle; that the notion of intensional structure makes good sense; that the content (i.e. inner content) of a thought or belief is determinate. Contents appear indeterminate only because our ascriptions of content are inaccurate. For arguments *contra* Quine, see Blackburn (1975), especially section II, and Churchland (1979), ch. 3.

[8] The 'basic idea' is, crudely, that A specifies the real, but inaccessible, contents of S's thoughts by modelling them on the semantic contents of publicly accessible sentences. A could not do this unless there already existed sentences available for use by A. If it were my concern in this paper to offer a theory of what it is for those sentences, or for sentences in general, to have semantic properties, my account would appeal to psychological facts about speakers and hearers (cf. Loar 1976b; Schiffer 1980).

I would assume a prior understanding of such notions as *belief* and *intention*. If I then wished to offer a general account of what it is for someone to have a belief or intention, including within this an explication of what it is for a belief or an intention to have content, it would be circular for me to appeal to semantic facts or to presuppose semantic notions.

But it is not at all the aim of this paper to give a theory of semantic content, nor even to provide a theory of mental content, in the sense of a general explication of what it is for a mental state to have content. The goal is to elucidate what it is to *specify* mental content—or rather, what it is for a particular ascriber A to specify the content of a thought in a particular individual S. The account I offer makes reference to sentences and their intensions. But no questions are begged, I hope. The sentences and intensions that I refer to are ontologically independent of S and S's thoughts. They are assumed to exist and to be available to A; A's world is already set up like that prior to S's entry on the scene.

[9] Can a human ascriber accurately specify the content of an animal's thought or belief, given that the animal's mind works very differently from a human mind? Clearly not, if he sticks to locutions like 'Fido believes that ——', 'Fido piensa que ——' etc, since content-specifications in *oratio obliqua* are egomorphic. This carries over equally to our ascriptions of *goals* to animals, and to the ways we describe goal-directed behaviour. (See Woodfield 1980.)

Stich (1979) made the point very clearly, but he exaggerated the consequences a little.

He suggested that the fact that we cannot express Fido's beliefs *accurately* casts doubts on the very idea of a dog's having beliefs. But perfectly accurate specification is an idealization, even in the case of fellow human beings (cf. my remarks on inaccuracy in the present paper, also see Stich's newer view (this volume)).

As far as second-order specifications are concerned, we must conclude that they too are inaccurate if they model the canine conceptual framework on to a *human* language. Still, we seem to be able to conceive of such a thing as a non-human language. We certainly believe that other animal species have systems of communication that are *like* languages (viz. whales, porpoises). If there did exist or could exist a language that accurately modelled the mind of a dog, human beings would doubtless be incapable of understanding it. That does not mean they would be unable to cite sentences of it. We are perhaps stretching the notion of 'specifying content by description' here! But we begin to see why there is no sharp dividing line between species that have minds and species that do not.

The second-order approach also illuminates why, or to what extent, it is impossible to judge nonsense. Wittgenstein in the *Tractatus* took this to be a test of the adequacy of a theory of judgement (cf. Wittgenstein 1921, 5.5422). If 'Nothing noths' had no intensional role whatever in English, then no thought could have a cognitive role analogous to the intension of 'Nothing noths' in English. (Actually, 'Nothing noths' is not totally devoid of intension in English; at least it has a grammatical structure, and it contains the word 'Nothing', so perhaps it is not complete and utter nonsense. The category of the nonsensical has no clear boundary.)

[10] Church's arguments were aimed at Carnap's 'sentential relation' analysis of belief-sentences (see Stich, this vol.). The 'Problem' does not arise for me as I am not offering an analysis. I am proposing an alternative to *oratio obliqua* which is superior for classificatory purposes. But the nub of Church's criticism is that Carnap tied the content of the belief too closely to a sentence. What it ought to be tied to is the intension that happens to be possessed by the sentence. (See Carnap 1956, sections 13, 14; Loar 1976b, pp. 146–7; Leeds 1979.)

[11] Here is how to view the celebrated 'arthritis' case from the second-order stance (I assume familiarity with Burge 1979a.)

(1) Both English and Twin-English are imperfect models of Sam and Twin-Sam (as we may call the two *Doppelgängers*). Burge states that Sam does not fully understand the English word 'arthritis', Twin-Sam does not fully understand the Twin-English word 'arthritis', There are four possible second-order specifications in play:

(a) Relative to a modelling of Sam on English, his thought-content is the intension of 'I have arthritis' in English.

(b) Relative to a modelling of Sam on Twin-English, his thought-content is the intension of 'I have arthritis' in Twin-English.

(c) Relative to a modelling of Twin-Sam on English, his thought-content is the intension of 'I have arthritis' in English.

(d) Relative to a modelling of Twin-Sam on Twin-English, his thought-content is the intension of 'I have arthritis' in Twin-English.

(2) Specification (a) is not perfectly accurate but it is the best we can get within the limits of the model. We accept it as true.

(3) Sam and Twin-Sam are physical replicas, so they have the same functional organization, and the same inner contents. It follows that (c) is the best specification we can get within the limits of a modelling of Twin-Sam on English.

(4) We are inclined to reject the statement 'Twin-Sam believes that he has arthritis'. We have been made vividly aware of the difference in meaning between 'arthritis (English)' and 'arthritis (Twin-English)'. Since (c) is the second-order equivalent of the

oratio obliqua ascription that we reject, we reject (c). We regard (d) as a superior specification.

(5) The question now is: what rational basis is there for accepting (a) while rejecting (c)? The model is the same, the systems modelled are the same. Evidently we *prefer* to model Sam on English and to model Twin-Sam on Twin-English. But *ex hypothesi* each language is an imperfect model, and each is imperfect *to the same degree for both subjects.*

(6) There is a reason for this preference, however. Even though Twin-English is not a better model than English of Twin-Sam at t, it is better policy for us to *use* Twin-English as the model of Twin-Sam. Native-language modelling generally offers the best long-term prospects, because S's conceptual structure is (normally) likely to approximate more closely to L_S as time goes by. We see this happening in Sam's case as soon as he corrects his belief in the light of the doctor's reply. We do not see it happening to Twin-Sam, but we can be reasonably sure that he would bring himself into line with Twin-Earthian practice if he discovered any mistake. On the assumption that Twin-English is likely to become a better model than it now is, we treat (c) as defeasible by (d).

Burge's own view is that the two replicas 'express different beliefs'—an ambiguous phrase, which on one reading implies that they *have* beliefs whose contents are in reality distinct. The falsity of that implication may be brought out with the help of an analogy. Imagine that you have two shoes that are precisely the same length, breadth and depth inside and out, one of which is in England and the other is left behind in Afghanistan. You know that the English one is a size 8 (U.K.), having checked with a shop-assistant; but the manufacturers skimped a little and it is fractionally below the norm for a size 8. Also, when you were in Afghanistan you took the other shoe to a shop and were told it was a size 293 (on the Afghan scale). Actually it was fractionally over the norm, but not enough over to count as a size 294. You now have before you a chart drawn up by the International Shoe Federation, which clearly shows that U.K. size 8 corresponds exactly to Afghan size 294. U.K. size 8 is not the same size as Afghan size 293! But this does not imply that your two shoes are in reality different in size. It would be false to conclude this, and also irrational for you to conclude it given that you know they *are* the same size. Had you *not* known antecedently that they were the same size however, it would have been not unreasonable to infer, from the evidence of the shopkeepers plus the chart, that they really differed.

Bibliography

ALSTON, W. P. 1966. 'Wants, Actions and Causal Explanations'. In *Intentionality, Minds and Perception*, ed. H.-N. Castañeda. Detroit: Wayne State U.P.

ANDERSON, J. R. 1976. *Languages, Memory and Thought*. Hillsdale N.J.: Lawrence Erlbaum Associates.

ANDERSON, J. R. and BOWER, G. 1973. *Human Associative Memory*. Washington D.C.: Hemisphere Publishing Co.

AQUILA, R. E. 1977. *Intentionality: A Study of Mental Acts*. Penn. State U.P., University Park and London.

AQUINAS, ST. THOMAS. *Summa Theologiae*. Blackfriars edition, 60 volumes, Eyre & Spottiswoode 1963–75.

ARMSTRONG, D. M. 1973. *Belief, Truth and Knowledge*. Cambridge: Cambridge U.P.

BACH, K. 1968. *Two Problems of Perception*. Ph.D. dissertation, University of California, Berkeley.

BACH, K. 1970. 'Part of What a Picture Is'. *British J. Aesthetics* 10.

BACH, K. 1978. 'A Representational Theory of Action'. *Philosophical Studies* 34.

BACH, K. forthcoming a. 'What's in a Name?' *Australasian Journal of Philosophy*.

BACH, K. forthcoming b. 'Referential/Attributive'. *Synthese*.

BACH, K. and HARNISH, R. M. 1979. *Linguistic Communication and Speech Acts*. Cambridge, Mass.: MIT Press.

BENNETT, J. 1976. *Linguistic Behaviour*. Cambridge: Cambridge U.P.

BLACKBURN, S. 1975. 'The Identity of Propositions'. In *Meaning, Reference and Necessity*, ed. S. Blackburn. Cambridge: Cambridge U.P.

BLACKBURN, S. 1979. 'Thoughts and Things'. *Aristotelian Society Supplementary Volume* LIII.

BLOCK, N. 1978. 'Troubles with Functionalism'. In *Perception and Cognition: Issues in the Foundations of Psychology: Minnesota Studies in the Philosophy of Science vol. IX*, ed. C. W. Savage. Minneapolis: University of Minnesota P.

BLOCK, N. forthcoming (ed.). *Readings in the Philosophy of Psychology*. Cambridge, Mass.: Harvard U.P.

BLOCK, N. and FODOR, J. 1972. 'What Psychological States Are Not' *Phil. Review* 81.

BOER, S. and LYCAN, W. 1975. 'Knowing Who'. *Philosophical Studies* 28.

BRANDT, R. B. and KIM, J. 1963. 'Wants as Explanation of Actions'. *J. Philosophy* LX.

BROAD, C. D. 1965. 'The Theory of Sensa'. In *Perceiving, Sensing and Knowing*, ed. R. J. Swartz. Garden City, N.Y.: Doubleday.

BURGE, T. 1977. 'Belief De Re'. *J. Philosophy* LXXIV.

BURGE, T. 1978. 'Belief and Synonymy'. *J. Philosophy* LXXV.

BURGE, T. 1979a. 'Individualism and the Mental'. In *Midwest Studies in Philosophy vol. IV: Studies in Metaphysics,* ed. P. A. French, T. E. Uehling and H. K. Wettstein. Minneapolis: U. of Minnesota P.

BURGE, T. 1979b. 'Sinning Against Frege'. *Phil. Review* 88.

CAMPBELL, D. forthcoming. 'Descriptive Epistemology: Psychological, Sociological and Evolutionary'. William James Lectures, Harvard University 1977.

CARNAP, R. 1956. *Meaning and Necessity.* 2nd edition, Chicago: Chicago U.P.

CASTAÑEDA, H.-N. 1966. '"He": A Study in The Logic of Self-Consciousness'. *Ratio* 8.

CASTAÑEDA, H.-N. 1967. 'Indicators and Quasi-Indicators'. *American Phil. Quarterly* 4.

CASTAÑEDA, H.-N. 1968. 'On the Logic of Attributions of Self-Knowledge to Others'. *J. Philosophy* LXV. 1968.

CHISHOLM, R. 1957. *Perceiving: A Philosophical Study.* Ithaca, N.Y.: Cornell U.P.

CHISHOLM, R. 1966. 'On Some Psychological Concepts and the "Logic" of Intentionality'. In *Intentionality, Minds and Perception*, ed. H.-N. Castañeda. Also 'Rejoinder' (to the commentary of Robert Sleigh on the above) in the same vol. Detroit: Wayne State U.P.

CHISHOLM, R. 1980. 'The Logic of Believing'. *Pacific Philosophical Quarterly* 61.

CHURCH, A. 1950. 'On Carnap's Analysis of Statements of Assertion and Belief'. *Analysis* 10.

CHURCHLAND, P. 1979. *Scientific Realism and The Plasticity of Mind.* Cambridge: Cambridge U.P.

DAVIDSON, D. 1963a. 'Actions, Reasons and Causes'. *J. Philosophy* LX.

DAVIDSON, D. 1963b. 'On the Method of Extension and Intension'. In *The Philosophy of Rudolf Carnap*, ed. P. A. Schilpp. La Salle, Illinois: Open Court.

DAVIDSON, D. 1967. 'Truth and Meaning'. *Synthese* 17.

DAVIDSON, D. 1968. 'On Saying That'. *Synthese* 19.

DAVIDSON, D. 1969. 'True to the Facts'. *J. Philosophy* LXVI.

DAVIDSON, D. 1973a. 'In Defense of Convention T'. in *Truth, Syntax and Modality*, ed. H. Leblanc. Amsterdam: North-Holland.

DAVIDSON, D. 1973b. 'Radical Interpretation'. *Dialectica* XXVII.

DAVIDSON, D. 1947a. 'On the Very Idea of a Conceptual Scheme'. *Proceedings and Addresses of the American Philosophical Association*, XLVII (1973/4).

DAVIDSON, D. 1974b. 'Psychology as Philosophy'. In *Philosophy of Psychology*, ed. S. C. Brown. London: Macmillan.

DAVIDSON, D. 1975. 'Thought and Talk'. In *Mind and Language*, ed. S. Guttenplan. Oxford: Oxford U.P.

DAVIDSON, D. 1977. 'Reality Without Reference'. *Dialectica* 31.

DAVIDSON, D. and HARMAN, G. 1972 (eds). *Semantics of Natural Language*. Boston: Reidel.

DENNETT, D. C. 1969. *Content and Consciousness*. London: Routledge and Kegan Paul.

DENNETT, D. C. 1973. 'Mechanism and Responsibility'. In *Essays on Freedom of Action*, ed. T. Honderich. London: Routledge and Kegan Paul. [Reprinted in Dennett 1978a.]

DENNETT, D. C. 1975. 'Brain Writing and Mind Reading'. In *Language, Mind and Knowledge: Minnesota Studies in the Philosophy of Science, vol. VII*, ed. K. Gunderson. Minneapolis: University of Minnesota P. [Reprinted in Dennett 1978a.]

DENNETT, D. C. 1977. 'Critical Notice: *The Language of Thought* by Jerry Fodor'. *Mind* LXXXVI. [Reprinted as 'A Cure for the Common Code?' in Dennett 1978a, ch. 6.]

DENNETT, D. C. 1978a. *Brainstorms*. Bradford Books, Montgomery, Vt., and Harvester Press, Sussex.

DENNETT, D. C. 1978b. 'Current Issues in the Philosophy of Mind'. *American Phil. Quarterly* 15.

DENNETT, D. C. forthcoming a. 'Three Kinds of Intentional Psychology'. Forthcoming in a Thyssen Foundation Philosophy Group volume, ed. Richard Healey.

DENNETT, D. C. forthcoming b. 'True Believers: The Intentional Strategy and Why it Works'. A 1979 Herbert Spencer Lecture, Oxford University.

DENNETT, D. C. unpublished. 'Quining Qualia'.

DESCARTES, R. 1641. *Meditations*. In *Philosophical Writings*, ed E. Anscombe and P. Geach. London: Nelson, 1954.

DONNELLAN, K. 1966. 'Reference and Definite Descriptions'. *Philosophical Review* 75.

DONNELLAN, K. 1968. 'Putting Humpty-Dumpty Together Again'. *Philosophical Review* 77.

DONNELLAN, K. 1970. 'Proper Names and Identifying Descriptions'. *Synthese* 21.

DONELLAN, K. 1974. 'Speaking of Nothing'. *Philosophical Review* 83.

DRETSKE, F. 1978. 'The Role of the Percept in Visual Cognition'. In *Perception and Cognition: Issues in the Foundations of Psychology: Minnesota Studies in the Philosophy of Science vol. IX*, ed. C. W. Savage. Minneapolis: University of Minnesota P.

DUMMETT, M. 1973. *Frege: Philosophy of Language.* London: Duckworth.

DUMMETT, M. 1975. 'What is a Theory of Meaning?' In *Mind and Language*, ed. S. Guttenplan. Oxford: Oxford U.P.

DUMMET, M. 1976. 'What is a Theory of Meaning? (II)'. In *Truth and Meaning*, ed. G. Evans and J. McDowell. Oxford: Clarendon Press.

DUMMETT, M. 1978. *Truth and Other Enigmas.* London: Duckworth.

DUMMETT, M. 1979. 'The Appeal to Use and The Theory of Meaning'. In *Meaning and Use*, ed. A. Margalit. Boston: Reidel.

EVANS, G. 1973. 'The Causal Theory of Names'. *Aristotelian Society Supplementary Volume* XLVII.

EVANS, G. 1980. 'Understanding Demonstratives'. In *Meaning and Understanding*, ed. H. Parret and J. Bouveresse. Walter de Gruyter, Berlin, New York.

EVANS, G. and McDOWELL, J. 1976 (eds.). *Truth and Meaning.* Oxford: Clarendon Press.

FIELD, H. 1972. 'Tarski's Theory of Truth'. *J. Philosophy* LXIX.

FIELD, H. 1977. 'Logic, Meaning and Conceptual Role'. *J. Philosophy* LXXIV.

FIELD, H. 1978. 'Mental Representation'. *Erkenntnis 13.*

FODOR, J. A. 1965. 'Explanations in Psychology'. In M. Black (ed.) *Philosophy in America.* Ithaca, N.Y.: Cornell U.P.

FODOR, J. A. 1968. *Psychological Explanation.* Random House, N.Y.

FODOR, J. A. 1975. *The Language of Thought.* Crowell, N.Y. and Harvester, Sussex.

FODOR, J. A. 1978. 'Propositional Attitudes'. *The Monist* 61.

FODOR, J. A. 1980. 'Methodological Solipsism Considered as a Research Strategy in Cognitive Psychology'. *The Behavioral and Brain Sciences* 3.

FREGE, G. 1892. 'On Sense and Reference'. In *Translations from the Philosophical Writings of Gottlob Frege*, ed. P. Geach and M. Black. Oxford: Blackwell, 1952.

FREGE, G. 1956. 'The Thought: A Logical Inquiry', trans. by A. M.

and Marcelle Quinton. *Mind* LXV. Reprinted in *Philosophical Logic*, ed. P. F. Strawson. Oxford: Oxford U.P., 1967.

GEACH, P. 1957. *Mental Acts*. London: Routledge and Kegan Paul.

GIBSON, J. J. 1966. *The Senses Considered as Perceptual Systems*. Houghton Mifflin, 1966; George Allen and Unwin, London, 1968.

GOLDMAN, A. I. 1970. *A Theory of Human Action*. Englewood Cliffs, N. J.: Prentice-Hall.

GOODMAN, N. 1961. 'About'. *Mind* LXXI.

GOODMAN, N. 1968. *Languages of Art*. Indianapolis: Bobbs-Merrill.

GOODMAN, N. 1978. *Ways of Worldmaking*. Indianapolis: Hackett.

GREGORY, R. 1966. *Eye and Brain*. London: Weidenfield and Nicolson; New York: McGraw-Hill.

HAMPSHIRE, S. 1975. *Freedom of The Individual*. Expanded edition, Princeton U.P.

HARMAN, G. 1970. 'Language Learning'. *Noûs* 4.

HARMAN, G. 1973. *Thought*. Princeton, N.J.: Princeton U.P.

HARMAN, G. 1974. 'Meaning and Semantics'. In *Semantics and Philosophy*, ed. M. Munitz and P. Unger. New York: NYU Press.

HARMAN, G. 1977. 'How to Use Propositions'. *American Phil. Quarterly* 14.

HAUGELAND, J. 1978. 'The Nature and Plausibility of Cognitivism'. *The Behavioral and Brain Sciences* 1.

HAUGELAND, J. 1979. 'Understanding Natural Language'. *J. Philosophy* LXXVI.

HINTIKKA, J. 1962. *Knowledge and Belief*. Ithaca, N.Y.: Cornell U.P.

HIRSCH, E. 1978. 'A Sense of Unity'. *J. Philosophy* LXXV.

HOFSTADTER, D. 1979. Gödel, Escher, Bach: An Eternal Golden Braid. Basic Books, N.Y. and Harvester Press, Sussex.

HORNSBY J. 1977. 'Singular Terms in Context of Propositional Attitude'. *Mind* LXXXVI.

HOUSE, W. unpublished. 'Charity and The World According to the Speaker'. Dissertation in progress, University of Pittsburgh.

JAMES, W. 1890. *Principles of Psychology*, vol. I. New York: Dover Publications, 1950.

KAPLAN, D. 1968. 'Quantifying In'. *Synthese* 19. [Also in *Words and Objections*, ed. D. Davidson and J. Hintikka. Dordrecht: Reidel, 1969.]

KAPLAN, D. 1973. 'Bob and Carol and Ted and Alice'. In *Approaches to Natural Language*, ed. J. Hintikka, J. Moravcsik and P. Suppes. Dordrecht: Reidel.

KAPLAN, D. 1978. 'Dthat'. In *Syntax and Semantics*, vol. 9, ed. Peter Cole. New York: Academic Press.

KAPLAN, D. forthcoming. 'Demonstratives'. The John Locke Lectures, Oxford University, 1980.

KATZ, J. 1978. 'The Theory of Semantic Representation'. *Erkenntnis* 13.

KATZ, J. and FODOR, J. 1963. 'The Structure of Semantic Theory'. *Language* 39.

KEMPSON, R. 1979. 'Presupposition, Opacity and Ambiguity'. In *Syntax and Semantics*, vol. 11, ed. Choon-Kyu Oh and David A. Dinnen. New York: Academic Press.

KIM, J. 1969. 'Events and their Descriptions: Some Considerations'. In *Essays in Honor of C.G. Hempel*, ed. N. Rescher *et al.* Dordrecht: Reidel.

KIM, J. 1976. 'Events as Property Exemplifications'. In *Action Theory*, ed. Myles Brand and Douglas Walton. Dordrecht: Reidel.

KIM, J. 1977. 'Perception and Reference without Causality'. *J. Philosophy* LXXIV.

KITELEY, M. (1968). 'Of What We Think'. *American Phil. Quarterly* 5.

KRIPKE, S. 1972. 'Naming and Necessity'. In *Semantics of Natural Language*, ed. D. Davidson and G. Harman. Boston: Reidel.

KRIPKE, S. 1977. 'Speaker's Reference and Semantic Reference'. In *Midwest Studies in Philosophy*, vol. II, ed. P. French *et al.* Minneapolis: U. of Minnesota P.

KRIPKE, S. 1979. 'A Puzzle About Belief'. In *Meaning and Use*, ed. A. Margalit. Boston: Reidel.

LEEDS, S. 1979. 'Church's Translation Argument'. *Canadian J. Phil.* IX.

LEWIS, D. 1966. 'An Argument for the Identity Theory'. *J. Philosophy* LXIII.

LEWIS, D. 1969. *Convention*. Cambridge, Mass.: Harvard U.P.

LEWIS, D. 1970. 'How to Define Theoretical Terms'. *J. Philosophy* LXVII.

LEWIS, D. 1972a. 'General Semantics'. In *Semantics of Natural Language*, ed. D. Davidson and G. Harman. Boston: Reidel.

LEWIS, D. 1972b. 'Psychophysical and Theoretical Identifications'. *Australasian J. Phil.* 50.

LEWIS, D. 1974. 'Radical Interpretation'. *Synthese* 27.

LEWIS, D. 1978. 'Truth in Fiction'. *American Phil. Quarterly* 15.

LEWIS, D. 1979. 'Attitudes *De Dicto* and *De Se*'. *Phil. Review* 87.

LOAR, B. 1972. 'Reference and Propositional Attitudes'. *Phil. Review* 80.

LOAR, B. 1976a. 'The Semantics of Singular Terms'. *Philosophical Studies* 30.

LOAR, B. 1976b. 'Two Theories of Meaning'. In *Truth and Meaning*, ed. G. Evans and J. McDowell. Oxford: Clarendon Press.

LOAR, B. 1980. 'Syntax, Functional Semantics, and Referential Semantics' (commentary on Fodor 1980). *The Behavioral and Brain Sciences* 3.

LOCKE, J. 1690. *Essay Concerning Human Understanding*. Clarendon Edition. Oxford U.P., 1975.

LURIA, A. R. 1972. *The Man With a Shattered World*. Harmondsworth, Middlesex: Penguin.

LYCAN, W. G. forthcoming. 'Functionalism and Psychological Laws'.

MARGALIT, A. 1979 (ed.). *Meaning and Use*. Boston: Reidel.

MARTIN, E. 1978. 'The Psychological Unreality of Quantificational Semantics'. In *Perception and Cognition: Issues in The Foundations of Psychology: Minnesota Studies in the Philosophy of Science, Vol. IX*, ed. C. W. Savage. Minneapolis: University of Minnesota P.

McCARTHY, J. and HAYNES, P. 1969. 'Some Philosophical Problems from the Standpoint of Artificial Intelligence'. In *Machine Intelligence* 4, ed. B. Meltzer and D. Michie. Edinburgh: Edinburgh U.P.

McDOWELL, J. 1977. 'On The Sense and Reference of a Proper Name'. *Mind* LXXXVI.

McDOWELL, J. 1980. 'Quotation and Saying That'. In *Reference, Truth and Reality*, ed. Mark Platts. London: Routledge and Kegan Paul.

McGINN, C. 1977. 'Charity, Interpretation and Belief'. *J. Philosophy* LXXIV.

McGINN, C. 1980. 'Philosophical Materialism'. *Synthese* 44.

MELDEN, A. I. 1961. *Free Action*. London: Routledge and Kegan Paul.

MINSKY, M. 1975. 'A Framework for Representing Knowledge'. In P. H. Winston (ed.) *The Psychology of Computer Vision*. New York: McGraw-Hill.

MORTON, A. 1975. 'Because He Thought He Had Insulted Him'. *J. Philosophy* LXXII.

NEISSER, U. 1976. *Cognition and Reality*. San Francisco: Freeman.

NELSON, R. J. 1978. 'Objects of Occasion Beliefs'. *Synthese* 39.

PEACOCKE, C. 1975. 'Proper Names, Reference and Rigid Designation'. In *Meaning, Reference and Necessity*, ed. S. Blackburn. Cambridge: Cambridge U.P.

PEACOCKE, C. 1976. 'Truth Definitions and Actual Languages'. In *Truth and Meaning*, ed. G. Evans and J. McDowell. Oxford: Clarendon Press.

PEACOCKE, C. 1979. *Holistic Explanation*. Oxford: Oxford U.P.

PERRY, J. 1977. 'Frege on Demonstratives'. *Phil. Review* 86.

PERRY, J. 1979. 'The Problem of the Essential Indexical'. *Noûs* 13.

PLATTS, M. 1980 (ed.). *Reference, Truth and Reality*. London: Routledge and Kegan Paul.

POLLOCK, J. L. 1980. 'Thinking about an Object'. *Midwest Studies in Philosophy*, vol. V.

PRIOR, A. N. 1976. *Papers in Logic and Ethics*, ed. P. Geach and A. Kenny. London: Duckworth.

PUTNAM, H. 1960. 'Minds and Machines'. In *Dimensions of Mind*, ed. S. Hook. New York: NYU Press. [Reprinted in Putnam 1975b.]

PUTNAM, H. 1966. 'The Mental Life of Some Machines'. In *Intentionality, Minds and Perception*, ed. H.-N. Castañeda. Detroit: Wayne State U.P. [Reprinted in Putnam 1975b.]

PUTNAM, H. 1967. 'The Nature of Mental States'. First published as 'Psychological Predicates' in *Art, Mind and Religion*, ed. Capitan and Merrill. U. of Pittsburgh P., 1967. [Reprinted in Putnam 1975b under new title.]

PUTNAM, H. 1974. 'Comment on Wilfrid Sellars'. *Synthese* 27.

PUTNAM, H. 1975a. 'The Meaning of "Meaning"'. In *Language, Mind and Knowledge: Minnesota Studies in the Philosophy of Science, vol. VII*, ed. Keith Gunderson. Minneapolis: University of Minnesota P. [Reprinted in Putnam 1975b.]

PUTNAM, H. 1975b. *Philosophical Papers Vol. II: Mind, Language and Reality*. Cambridge: Cambridge U.P.

PUTNAM, H. 1978. *Meaning and The Moral Sciences*. London: Routledge and Kegan Paul.

PYLYSHYN, Z. 1979. 'Complexity and the Study of Artificial and Human Intelligence'. In *Philosophical Perspectives in Artificial Intelligence*, ed. M. Ringle. Humanities Press, Atlantic Highlands, N.J. and Harvester, Sussex.

QUINE, W. V. O. 1956. 'Quantifiers and Propositional Attitudes'. *J. Philosophy* LIII. [Reprinted in Quine: *The Ways of Paradox*. New York: Random House, 1966.]

QUINE, W. V. O. 1960. *Word and Object*. Cambridge, Mass.: MIT Press.

QUINE, W. V. O. 1969. 'Propositional Objects'. In *Ontological Relativity and Other Essays*. New York: Columbia U.P.

QUINE, W. V. O. 1979. 'Use and its Place in Meaning'. In *Meaning and Use*, ed. A. Margalit. Boston: Reidel.

REDER, L. M. 1976. *The Role of Elaborations in The Processing of Prose*. Ph.D. Dissertation, University of Michigan (Department of Psychology).

RORTY, R. 1972. 'The World Well Lost'. *J. Philosophy* LXIX.

ROSCH, E. 1973a. 'Natural Categories'. *Cognitive Psychology* 4.

ROSCH, E. 1973b. 'On the Internal Structure of Perceptual and Semantic Categories'. In *Cognitive Development and the Acquisition of Language*, ed. T. Moore. New York: Academic Press.

ROSCH, E. 1975. 'Cognitive Representation of Semantic Categories'. *J. Experimental Psychology: General* 104.

ROSCH, E. 1977. 'Classification of Real-World Objects: Origins and Representation in Cognition'. In *Thinking*, ed. P. N. Johnson-Laird and P. C. Wason. Cambridge: Cambridge U.P.

ROSCH, E, and MERVIS, C. B. 1975. 'Family Resemblances: Studies in the Internal Structure of Categories'. *Cognitive Psychology* 7.

RUSSELL, B. 1940. *An Inquiry into Meaning and Truth.* New York: Norton.

RUSSELL, B. 1959. *Mysticism and Logic.* London: Allen & Unwin.

SCHANK, R. and ABELSON, R. 1977. *Scripts, Plans, Goals and Understanding.* Hillsdale, N.J.: John Wiley and Sons.

SCHIFFER, S. 1973. *Meaning.* Oxford: Oxford U.P.

SCHIFFER, S. 1977. 'Naming and Knowing'. In *Midwest Studies in Philosophy*, vol. II, ed. P. French *et al.* Minneapolis: University of Minnesota P.

SCHIFFER, S. 1978. 'The Basis of Reference'. *Erkenntnis* 13.

SCHIFFER, S. 1980. 'Truth and the Theory of Content'. In *Meaning and Understanding*, ed. H. Parret and J. Bouveresse. Walter de Gruyter, Berlin, New York.

SEARLE, J. R. 1979a. 'What is an Intentional State?' *Mind* LXXXVIII.

SEARLE, J. 1979b. 'Referential and Attributive'. *The Monist* 62. [Also in Searle: *Expression and Meaning.* Cambridge: Cambridge U.P. 1980.]

SELLARS, W. 1956. 'Empiricism and The Philosophy of Mind'. In *Minnesota Studies in The Philosophy of Science*, vol. I, ed. H. Feigl and M. Scriven. Minneapolis: University of Minnesota P. [Reprinted in Sellars: *Science, Perception and Reality.* London: Routledge and Kegan Paul, 1963.]

SELLARS, W. 1967. *Philosophical Perspectives.* Springfield, Illinois; Charles C. Thomas.

SELLARS, W. 1974. 'Meaning as Functional Classification'. *Synthese* 27.

SLOMAN, A. 1978. *The Computer Revolution in Philosophy: Philosophy, Science and Models of Mind.* Humanities Press, Atlantic Highlands, N.J. and Harvester, Sussex.

SOSA, E. 1970. 'Propositional Attitudes *De Dicto* and *De Re*'. *J. Philosophy* LXVII.

STALNAKER, R. 1976. 'Propositions'. In *Issues in The Philosophy of*

Language. ed. Alfred MacKay and Daniel Merrill. New Haven: Yale U.P.

STICH, S. P. 1976. 'Davidson's Semantic Program'. *Canadian J. Phil.* 4.

STICH, S. P. 1978a. 'Beliefs and Sub-Doxastic States'. *Phil. of Science* 45.

STICH, S. 1978b. 'Autonomous Psychology and the Belief-Desire Thesis'. *The Monist* 61.

STICH, S. P. 1979. 'Do Animals Have Beliefs?' *Australasian J. Of Phil.* 57.

STICH, S. P. 1980. 'Paying the Price for Methodological Solipsism'. *The Behavioral and Brain Sciences* 3.

STICH, S. P. forthcoming. 'Computation without Representation'. To appear in *The Behavioral and Brain Sciences.*

STRAWSON, P. F. 1959. *Individuals.* London: Methuen.

STROUD, B. 1968. 'Conventionalism and The Indeterminacy of Translation'. *Synthese* 19.

TODD, W. 1977. 'The Use of Simulations in Analytic Philosophy'. *Metaphilosophy* 8.

TVERSKY, A. 1977. 'Features of Similarity'. *Psychological Review* 84.4.

ULLIAN, J. and GOODMAN, N. 1977. 'Truth About Jones'. *J. Philosophy* LXXIV.

VENDLER, Z. 1976. 'Thinking of Individuals'. *Noûs* 10.

VENDLER, Z. forthcoming. 'Reference and Introduction', delivered at Oxford University, Spring 1979.

WALLACE, J. 1972. 'Belief and Satisfaction'. *Noûs* 6.

WASON, P. C. and JOHNSON-LAIRD, P. N. 1972. *Psychology of Reasoning: Structure and Content.* London: B. T. Batsford.

WIMSATT, W. C. 1974. 'Complexity and Organization'. In *PSA 1972* (Philosophy of Science Association), ed. K. Schaffner and R. S. Cohen. Dordrecht: Reidel.

WINOGRAD, T. 1972. *Understanding Natural Language.* Edinburgh: Edinburgh U.P.

WITTGENSTEIN, L. (1921). *Tractatus Logico-Philosophicus.* Trans. and ed. D. F. Pears and B. F. McGuinness. London: Routledge and Kegan Paul, 1961.

WOODFIELD, A. 1976. *Teleology.* Cambridge: Cambridge U.P.

WOODFIELD, A. 1980. 'Some Connections between Ascriptions of Goals and Assumptions of Adaptiveness'. In *Logic, Methodology and Philosophy of Science, VI,* ed. L. J. Cohen, J. Los, H. Pfeiffer and K.-P. Podewski. Amsterdam: North-Holland.

WOODS, W. 1975. 'What's in a Link'. In *Representation and Understanding,* ed. D. Bobrow and A. Collins. New York: Academic Press.

WOODS, W. forthcoming. 'Procedural Semantics as a Theory of

Meaning'. In *Linguistic Structure and Discourse Setting*, ed. Joshi, Sag and Webber. Cambridge: Cambridge U.P.

ZEMACH, E. M. 1976. 'Putnam's Theory on the Reference of Substance Terms'. *J. Philosophy* LXXIII.

ZEMAN, J. 1963. 'Information and the Brain'. In *Nerve, Brain and Memory Models: Progress in Brain Research, vol. II*, ed. N. Weiner and J. P. Schadé. Amsterdam and N.Y.: Elsevier Publishing Co.

Index of Names

Subject Index

ATC its taken
as data —
reference point
role?

do they need to track?

Trading cardinint, not
very.